T0311833

Financial Modelling in Commodity Markets

CHAPMAN & HALL/CRC
Financial Mathematics Series

Aims and scope:
The field of financial mathematics forms an ever-expanding slice of the financial sector. This series aims to capture new developments and summarize what is known over the whole spectrum of this field. It will include a broad range of textbooks, reference works and handbooks that are meant to appeal to both academics and practitioners. The inclusion of numerical code and concrete real-world examples is highly encouraged.

Series Editors

M.A.H. Dempster
Centre for Financial Research
Department of Pure Mathematics and Statistics
University of Cambridge

Dilip B. Madan
Robert H. Smith School of Business
University of Maryland

Rama Cont
Department of Mathematics
Imperial College

Derivative Pricing
A Problem-Based Primer
Ambrose Lo

Portfolio Rebalancing
Edward E. Qian

Interest Rate Modeling
Theory and Practice, 2nd Edition
Lixin Wu

Metamodeling for Variable Annuities
Guojun Gan and Emiliano A. Valdez

Modeling Fixed Income Securities and Interest Rate Options
Robert A. Jarrow

Financial Modelling in Commodity Markets
Viviana Fanelli

For more information about this series please visit:
https://www.crcpress.com/Chapman-and-HallCRC-Financial-Mathematics-
Series/book-series/CHFINANCMTH

Financial Modelling in Commodity Markets

Viviana Fanelli
University of Bari Aldo Moro

CRC Press
Taylor & Francis Group
Boca Raton London New York

CRC Press is an imprint of the
Taylor & Francis Group, an **informa** business

A CHAPMAN & HALL BOOK

CRC Press
Taylor & Francis Group
6000 Broken Sound Parkway NW, Suite 300
Boca Raton, FL 33487-2742

International Standard Book Number-13: 978-1-138-73910-9 (Hardback)
978-0-367-44286-6 (Paperback)

Visit the Taylor & Francis Web site at
http://www.taylorandfrancis.com

and the CRC Press Web site at
http://www.crcpress.com

To my parents and to my husband, Marco, and our beloved son, Manuele.

Contents

Preface

The liberalization of commodity markets gave rise to new patterns of financial product prices and the need for models that could really describe price dynamics grew exponentially, above all in order to improve decision making for all of the agents involved in commodity issues. Academics and practitioners, given the intrinsic risk-return characteristics of commodities, currently treat them as a particular asset class which has led to a spectacular growth in spot and derivative trading. Commodity asset class is therefore of interest for investors that includes hedge funds as well as trading companies.

The aim of this book is to provide a basic and self-contained introduction to the ideas underpinning financial modelling of products in commodity markets. The book offers a concise and operational vision of the main models used to represent, assess and simulate real assets and financial positions related to the commodity markets. It discusses statistical and mathematical tools important for estimating, implementing and calibrating quantitative models used for pricing and trading commodity-linked products and for managing basic and complex portfolio risks.

Hence, readers of the proposed book would include master's degree students and Ph.D. students from quantitatively-oriented fields of the economics and finance, as well as mathematics, engineering and statistics; and it would include scientists and research workers from either academic or industrial settings, where data analysis and financial models are performed; and finally it could be read by anyone wishing to make that first read into the subject.

The book will be aimed for practitioners and theorists alike. For the practitioner, often interested in model building and analysis, we provide the cornerstone ideas. Any pricing model must be well motivated and we will discuss what the objectives for such models are. Again, starting from the basic ideas we guide the reader through the development and building up of a model from the initial desire to analyse data to the completion of the task. Furthermore, we propose some interesting practical examples of real situations which an investor (individual or company) can be involved in.

Introduction

This book reflects the needs of a reader who requires a synthetic guide to the definition and implementation of models for pricing commodity-linked products and their use. Therefore, on the one hand the author summarises the main steps of the financial-mathematical modelling, on the other hand she does not fail to highlight the main features of the models and to give useful insights for their implementation in Excel environment. The practical exercises are necessary to make the reader understand how it is possible to apply all the described theory in real-life situations.

The first chapter deals with the most common and used financial products in commodity markets. In particular, it contains the explanation of their payoffs and the description of the main characteristics for operational scopes.

The second and third chapters contain the most relevant models respectively for the spot and forward/futures prices. They are very concise in providing the main features of such models and giving some hints for their implementation through the Excel software.

The main topic of Chapter 4 is the pricing of financial options. In particular, options, which have the spot price as underlying, are priced through the continuous time model of Black and Scholes and the discrete time Binomial model. If the underlying is the forward/futures price, we propose the Black model. Furthermore, the Monte Carlo approach to price other financial derivatives is introduced.

Chapter 5 contains some complete applications of the theory explained in the previous chapters. The reader will be able to identify with a risk manager who needs to hedge the electricity price risk when he buys electricity from a producer and sells it to the final consumer; with a financial operator who needs to choose the best model to describe the evolution of the spark spread; and with a trader who develops strategies to trade a portfolio of crude oils.

Finally, Chapter 6 recalls some basic statistical tools used in the book.

Chapter 1

Commodity-linked Products

1.1 Forward Contracts and Exchange Traded Futures

1.1.1 Forward Price

A forward contracts is a bilateral agreement to purchase or sell a certain amount of a commodity on a fixed future date at a predetermined contract price. The fixed future date is the delivery date of the commodity, and the predetermined contract price is the forward price. Thus, the participants of a forward contract on a commodity lock in a price today for future delivery. In Figure 1.1 we show the role of forward contract participants. The main characteristics of forward contracts are:

- They are over-the-counter (OTC) trades, executed through brokers.

- There are no cash flows until delivery.

- On delivery date the seller of the contract has the obligation to deliver the commodity in return for the forward price.

FIGURE 1.1: Forward contract participants

- A forward contract involves credit risk, where one of the counterparties does not, or cannot, fulfil his obligation to deliver or pay the commodity.

A forward contract can be mainly used for:

1. hedging the obligation to deliver or purchase a commodity at a future date;

2. guaranteeing a sales profit from a commodity production;

3. speculating on rising or falling commodity prices in case there is no liquid futures market.

Consider a customer who is long on forward contract for delivery date T at contract price K. The payoff at time T is given by the difference between the actual price $S(T)$ of the underlying and the contract forward price K. It is illustrated in Figure 1.2.

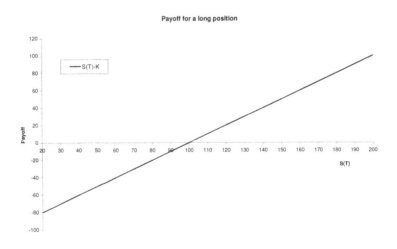

FIGURE 1.2: The payoff at maturity of a forward contract

TABLE 1.1: Forward contract strategy

	t	T
Long Forward at time 0	$X(t)$	$S(T) - K$
Short Forward at time t	0	$F(t,T) - S(T)$
Cash Flow	$X(t)$	$F(t,T) - K$

Let us assume that at an initial time 0 a customer sells a forward contract with delivery T and contract price K. The forward price will be $F(0,T)$. We want to estimate the value of the contract, $X(t)$, at a generic time t. In order to do this, at time t we sell another forward contract at the current market forward price $F(t,T)$, so that the physical deliveries of both forward contracts in the portfolio cancel.

In fact, if we indicate with $S(T)$ the market price of the underlying forward contract at delivery T, Table 1.1 illustrates the strategy payoff at every considered time.

In order to avoid arbitrage possibilities, the fair value of a forward contract $X(t)$ is uniquely given by

$$X(t) = e^{-r(T-t)}(F(t,T) - K),$$

where r is the effective risk-free interest rate.

Example 1 *Example of Fair Value.*
An electricity producer buys 10,000t coal to be delivered at time T at a price $K = 50USD/t$. At time t, the forward price for coal to be delivered at time T is $F(t,T) = 60USD/t$.

$$K = 50USD/t \qquad F(t,T) = 60USD/t$$

0	$t = 1$	$T = 2$
	$\uparrow$	
	$X(t) = ?$	

Assuming that at time t the current interest rate is 4%, the fair value of the forward contract is

$$X(t) = \frac{1}{1.04} \cdot 10,000 \cdot (60 - 50) = 96154 \quad USD$$

Some commodities, as for example electricity, have to be delivered over a period of time instead of a single day in the future. Then, the forward contract has a sequence of delivery dates $T_1, ..., T_n$, so it can be decomposed into n forward contracts so that each one has a single delivery date. If K is the fixed forward contract price, the present fair value of the contract at a generic time t is given by

$$X(t) = \sum_{i=1}^{n} e^{-r(T_i-t)}(F(t,T_i) - K). \qquad (1.1)$$

The fair price of the forward contract is gotten by solving $X(t) = 0$ in (1.1), so that

$$K = \frac{\sum_{i=1}^{n} e^{-r(T_i - t)} F(t, T_i)}{\sum_{i=1}^{n} e^{-r(T_i - t)}}.$$

For delivery over a continuous period $[T_1, T_2]$, the fair value formulas are

$$X(t) = \int_{T_1}^{T_2} e^{-r(T-t)} (F(t,T) - K) dT \quad \Rightarrow \quad K = \frac{\int_{T_1}^{T_2} e^{-r(T-t)} F(t,T) dT}{\int_{T_1}^{T_2} e^{-r(T-t)} dT}, \forall t \leq T_1.$$

1.1.2   Futures Price

A futures contract is a standardised forward contract traded at a commodity exchange. As for the forward contract, the futures contract is an agreement to buy or sell a certain amount of a commodity on a fixed future date at a predetermined contract price. However, the participants in the contract are three: the buyer of the contract, the seller and the clearing house. The clearing house serves as the central counterparty for all transactions. Trading participants usually pay an initial margin as a guarantee. At each trading day the futures contract is updated by determining the settlement price and gains or losses are immediately realised at a margin account, called variation margin, to the clearing house. In this way, there is not a counterparty risk because no unrealised losses may occur that could impose a substantial credit risk. Futures contracts often are settled financially and they do not lead to physical delivery. In Figure 1.3 we represent an example of hedging the obligation to sell a commodity at a future date. A crude oil producer wants to hedge the price of July future crude oil production with a futures. In particular, he locks in the July price by selling a July futures contract with delivery price $120. At the delivery date in July the crude oil market price is $113.50. Because the producer does not want to make delivery of the futures contract, he buys back the July futures contract at the prevailing market price to close out his position. In this way he would incur a gain of $6.5 on the futures contract.

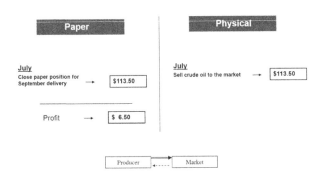

FIGURE 1.3: Futures example

We provide a list of four main differences between forward and futures contracts:

- Futures contracts are exchange-traded and, therefore, are standardised contracts; forward contracts are private agreements between two parties who establish terms and conditions;

- Forward contracts provide for credit risk, in particular counterparty risk; on the other hand futures contracts are credit risk free;

- Futures contracts are marked to market daily and the settlement occurs over a certain number of dates; instead forward contracts are settled on delivery at the end of the contract;

- Forward price and futures price have the same value only when the interest rate is constant over time and independent from the underlying price.

1.1.3 Spot-forward Relationship

The spot-forward relationship can be described by distinguishing between storable commodities and nonstorable commodities.

A commodity is stored only if the present value of selling at a future time T is at least as great as that of selling today. If $F(0,T)$ is today's price of a forward with delivery time T and $S(0)$ is the current spot price of the commodity, the following relationship holds:

$$F(0,T) \geq S(0)e^{rT} + \lambda(0,T),$$

where $\lambda(0,T)$ is the storage cost for one unit of commodity from time 0 to T and r is the current interest rate. There is no difference between selling today or at time T if

$$S(0) = e^{-rT}[F(0,T) - \lambda(0,T)].$$

When the storage costs are paid continuously at a rate λ, we have

$$F(0,T) = S(0)e^{(r+\lambda)T}.$$

We observe that storage cost is like a negative dividend! It affects the forward price. In fact, the forward curve can rise faster than the interest rate because the selling price must compensate the commodity merchant for both the financial cost of storage (interest) and the physical cost of storage.

Example 2 *What is the February Forward Price of wheat that is stored from November to February?*

- *November price of wheat: $4/bushel*

- *Monthly interest rate: 1%*

- *Storage costs: $0.05/month per bushel*
  Future value of storage costs=
  $4 + (4 \cdot 1.01) + (4 \cdot 1.01^2) = (4/0.01) \cdot [(1+0.01)^3 - 1] =$
  $0.1515

TABLE 1.2: Reverse cash-and-carry strategy

Transaction	Time 0	Time 1
Long forward at $F(0,T)$	0	$S(T) - F(0,T)$
Short $e^{-\delta}$ commodity units with rate δ	$+S(0)e^{-\delta T}$	$-S(T)$
Lend at r	$-S(0)e^{-\delta T}$	$+S(0)e^{(r-\delta)T}$
Total	0	$S(0)e^{(r-\delta)T} - F(0,T)$

$$Price = 4 \cdot (1.01)^3 + 0.1515 = 4.272.$$

In addition to storage costs, commodity prices also incorporate the so-called convenience yield. The convenience yield measures the monetary convenience for holding commodities. If c is the continuously compounded convenience yield, proportional to the value of the commodity, we have that:

- Commodity lender saves $\lambda - c$ by not physically storing the commodity

- Commodity borrower pays $c - \lambda$ by physically storing the commodity

In the case of the presence of convenience yield, the commodity price can be obtained by considering the reverse cash-and-carry strategy in Table 1.2.

In order to avoid arbitrage opportunities should be $F(0,T) = S(0)e^{(r+\lambda-c)T}$. However, the real no-arbitrage pricing region is $S(0)e^{(r+\lambda-c)T} \leq F(0,T) \leq S(0)e^{(r+\lambda)T}$.

We could summarise the characteristics of the convenience yield as follows:

- It is hard to observe;

- It depends on the inventories and reflects expectations about the availability of commodities (immediacy of the market);

- It explains the patterns in storage;

- It provides additional parameters to better explain the forward curve.

A nonstorable commodity, like electricity, must be consumed when it is produced. The price is set by demand and supply at a point in time. Price swings over the day primarily represent changes in the expected spot price, which in turn reflects changes in demand over the day. Then, the forward price reveals information about the future price of the commodity, as it was a price discovery. The forward price can be obtained by implementing the following strategy. Let us consider a portfolio consisting of a long position in a commodity forward contract at price $F(0,T)$ and a zero-coupon bond that pays $F(0,T)$ at time T. Strategy payoffs at the current interest rate r are illustrated in the Table 1.3.

The portfolio represents a synthetic commodity because it has the same value as a unit of the commodity at time T and its price is the discounted forward price.

TABLE 1.3: Forward-bond strategy

Position	Time 0	Time T
Long forward at $F(0,T)$	0	$S(T) - F(0,T)$
Short Bond with nominal value $F(0,T)$	$-e^{-rT}F(0,T)$	$F(0,T)$
Cash Flow	$-e^{-rT}F(0,T)$	$S(T).$

Therefore, we can apply no-arbitrage conditions and determine the forward price by using the expected spot price discounted at an appropriate discount rate d:

$$e^{-rT}F(0,T) = E_0[S(T)]e^{-dT},$$
$$e^{-rT}F(0,T) = E_0[S(T)]e^{-dT},$$
$$F(0,T) = E_0[S(T)]e^{(r-d)T},$$

where $E_0[\cdot]$ is the expected value given all information available at time 0. We conclude that the time T forward price discounted at the risk-free rate r back to time 0 is the present value of a unit of commodity received at time T and $(r-d)$ is the risk premium on the commodity. Then, the expected spot price and the forward price differ only on the risk premium and the estimation of the risk premium must be precise in order to have a reliable expected spot price. Often, the expected spot rate $E_0(S(T))$ is substituted by the term $S(0)e^{gT}$, where $S(0)$ is the spot price and g is the commodity price expected growth rate, which could be given by analysts' forecasts, for example. Therefore, the forward price is $F(0,T) = S(0)e^{(r-\alpha)T}$, where $\alpha = d - g$. The rate α could be seen as the negative of the storage cost ($\alpha = -\lambda$): the commodity borrower virtually stores the commodity for the lender, who receives it back at a future date. The lender pays a negative dividend on the amount due to the borrower[1].

1.1.3.1 Spot/Futures Arbitrage and the Basis

The spot price of the commodity underlying the futures contract should converge to the futures price at its delivery point. On the contrary, usually the futures price is greater than the spot price. This is because the underlying of the futures contract has a similar but not identical source of risk as in the spot market. In addition, at the maturity of the futures contract, the seller of the commodity has many externalities, such as grade of the commodity, location, chemical attributes, that affect the final price.

Therefore, an arbitrage occurs when purchasing spot commodity and selling futures contracts simultaneously. The opportunity of arbitrage vanishes quickly when the equilibrium between markets is restored. In fact, when the demand in the spot market rises, the spot price increases and the forward supply as well. Consequently, the forward price decreases until it is less than the spot price plus the carrying

---

[1] See McDonald [54] for a more detailed study.

Time 0:

Buy spot jet fuel: -$1.90 per gallon

Sell one-month futures contracts

Time 1 month:

Cost of storing: -$.02 per gallon

Two options:

- Satisfy the short futures contract   +$1.93 per gallon
- Offset the futures contract at profit and sell in the spot market

Profit: $0.01 per gallon

FIGURE 1.4: Spot/futures arbitrage

charges. The futures markets contribute greatly to the efficient rationing of goods over time and smooth flow of goods from producers to ultimate consumers.

Example 3 *Let us consider an airplane company that needs jet fuel in one month. We assume:*

- *Spot price: $ 1.90 per gallon;*

- *Costs to finance and store: $0.02 per month;*

- *One-month futures contract price: $1.93 per gallon;*

The company can earn profit implementing the arbitrage strategy in Figure 1.4.

The difference between the futures price and the spot price is called "the basis". If we consider only the spot price at the delivery point of the futures price, we can distinguish three components of the basis. They are shown in Figure 1.5.

Example 4 *Let us consider the gasoline markets:*

- *Futures price in Italy: $ 1.87 per liter;*

- *Spot price in Italy: $1.83 per liter;*

- *Spot price in France: $1.77 per liter;*

Basis= 1.87-1.83+1.83-1.77=$0.1 per liter.

On the contrary, in the Netherlands the spot price is $2.02, so the basis is negative.

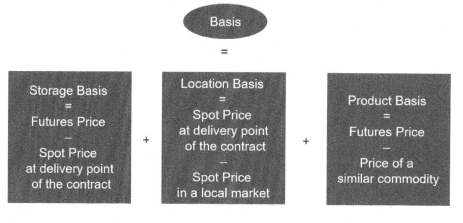

FIGURE 1.5: Basis components

1.2 Options

1.2.1 European Options

European financial options are called plain vanilla options because they represent the basic financial derivatives that provide for a buying or selling option. The name plain vanilla derives from the idea that the basic flavour of the ice cream is vanilla. A call option contract gives

- to the option holder the *right*, but not the obligation, to purchase a certain commodity at a predetermined strike price K at a maturity point T.

- to the option seller the *obligation* to deliver the commodity upon exercise by the option holder.

A put option contract gives

- to the option holder the *right*, but not the obligation, to sell a certain commodity at a predetermined strike price K.

- to the option seller the *obligation* to purchase the commodity upon exercise by the option holder.

The option seller receives an option premium at the time the contract is signed and options can be traded as over-the-counter products or through a commodity exchange. In Figures 1.6 and 1.7 we show the payoff at maturity T of both option types. The solid line represents the buyer payoff, the dashed line the writer payoff.

We highlight that at maturity the writer of the options received the payoff of the options with a negative sign, namely $-\max(S(T)-K,0)$ for the call and $-\max(K-S(T),0)$ for the put.

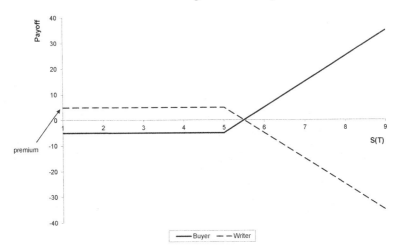

FIGURE 1.6: Call option payoff = $\max(S(T) - K, 0)$

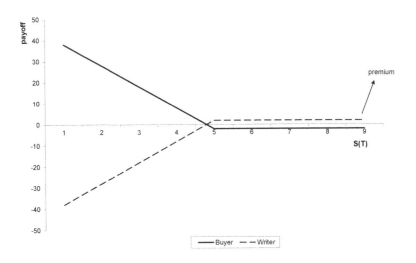

FIGURE 1.7: Put option payoff = $\max(K - S(T), 0)$

There exists a relationship between a call and a put option written on the same underlying with the same maturity. It is a formula called put-call parity that allows to calculate the European call option price given the put price and vice versa. We can obtain the relationship implementing the following strategy. Consider a portfolio consisting of:

- a long call option with strike K and maturity T and whose fair value $c_{K,T}$,

- a short put option with strike K and maturity T and whose fair value $p_{K,T}$.

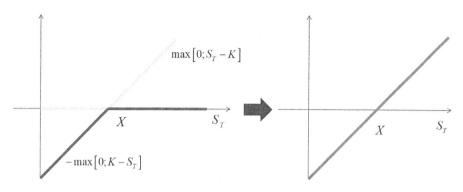

FIGURE 1.8: Put-call strategy

At maturity T the payoff of the portfolio is

$$\max(0, S(T) - K) - \max(0, K - S(T)) = \begin{cases} 0 - (K - S(T)), & \text{if } S(T) < K; \\ S(T) - K - 0, & \text{if } S(T) \geq K. \end{cases}$$

In both cases the payoff is $S(T) - K$ which is also the payoff of a forward contract with strike K, see Figure 1.8.

Therefore, a portfolio of long call and short put replicates a forward contract, such that at time t the following put-call parity relationship holds

$$c_{K,T} + p_{K,T} = e^{-rT}(S(T) - K).$$

1.2.2 American Options

An American option is a path-dependent option, that means the payoff is linked to the behaviour of the underlying over the lifetime of the option. In fact the right of the option holder can be exercised at any time τ during the lifetime $[t, T]$ of the option. It is obviously more expensive than a European option. The payoff at time $\tau \in [t, T]$ is $\max(0, S(\tau) - K)$ and in case of physical settlement of the option, the price $S(\tau)$ represents the value of the commodity being delivered. Calculating an American option price is more complex than evaluating a European option. We contemplate the Monte Carlo method in Section 4.4. However other approaches exist, as for example the implementation of an optimal exercise strategy. It consists of finding the optimal decision for each market price situation whether to exercise or not exercise.

1.2.3 Option Strategies

Any financial operator can arrange strategies by combining short and/or long positions on the underlying asset and/or options in order to hedge price risk or speculate. The diagrams of strategies that involve a position on the underlying asset and an option are shown in Figure 1.9. The dotted lines represent the payoffs of the position on the underlying asset and on the option; the solid line indicates the strategy profit.

Other strategies are illustrated in the following list.

- A cap is a strip of call options with the same strike price; whereas a floor is a strip of put options;

- Bull spread: a long call with strike price K_1 and a short call with strike price K_2, where $K_2 > K_1$, see Figure 1.10;

- Bear spread: a long call with strike price K_2 and a short call with strike price K_1, where $K_2 > K_1$, see Figure 1.11;

- Butterfly spread: a long call with strike price K_1, a long call with strike price K_3, two short calls with strike price K_2 where $K_1 < K_2 < K_3$, see Figure 1.12;

- Straddle: a long call and a long put with strike prices K, see Figure 1.13;

- Strip: a long call and two long puts with strike prices K; Strap: a long put and two long calls with strike prices K, see Figure 1.14;

- Strangle: a long call with strike price K_2 and a long put with strike price K_1, where $K_2 > K_1$, see Figure 1.15;

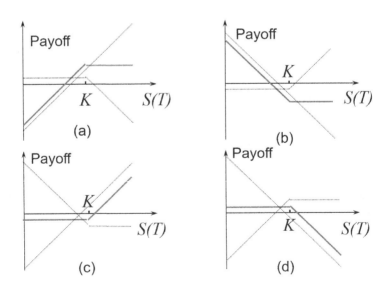

FIGURE 1.9: Option strategies: (a) long asset, short call, (b) short asset, long call, (c) long asset, long put, (d) short asset, short put

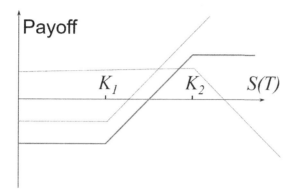

FIGURE 1.10: Bull spread payoff

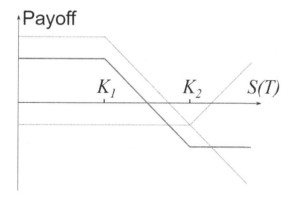

FIGURE 1.11: Bear spread payoff

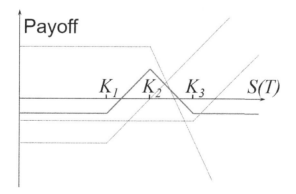

FIGURE 1.12: Butterfly spread payoff

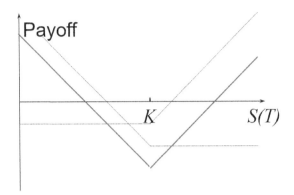

FIGURE 1.13: Straddle payoff

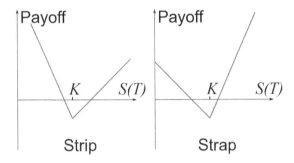

FIGURE 1.14: Strip and strap payoffs

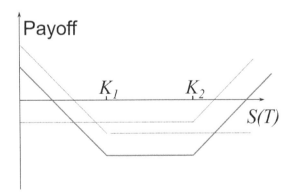

FIGURE 1.15: Strangle payoff

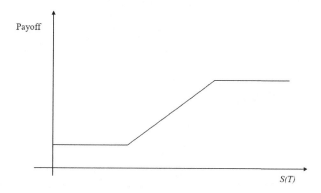

FIGURE 1.16: Costless collar payoff

We highlight that if we refer to Figure 1.9 and joint strategy (a), called covered call, and strategy (c), called protective put, we have a collar. Usually it is a costless strategy and its payoff is illustrated in Figure 1.16. It is used, for example, by a producer who is concerned about falling market prices. He could simply buy a floor, but the premium may be prohibitive. If he is prepared to sacrifice some of his upside he could reduce, or even eliminate the cost of the strategy altogether. It is possible to produce zero-cost collars by a smart choice of strike levels of the put and the call.

1.2.4 Exotic Options

Exotic options are modifications to the classical plain vanilla options. The payoff is path-dependent. This means that it depends on the underlying price trajectory during the entire life of the option, or some part of it[2]. Furthermore, many exotic options have two or more commodities as underlying, so that the payoffs depend on the prices of more commodities. We speak about multiple-commodity options. In Table 1.4 we list the main options with their characteristics and we indicate some common applications[3].

1.3 Swaps

Commodity swaps are mid- and long-term risk management instruments used to lock in a fixed price for a commodity over a specific time period.

---

[2]We recall that in case of European option, the payoff is calculated only at the expiration of the option, that is the payoff depends on the price of the underlying at exercise.

[3]We suggest you refer to Kaminski [47] for a more exhaustive discussion of the topic, and to Fanelli and Ryden [30] for an example of exotic option. In addition, see Fanelli et al. [33] for the pricing of an Asian option in the electricity market.

TABLE 1.4: Option types

Instrument	Payoff	Example
Asian option	For a call, $\max[\bar{S} - K; 0]$; for a put, $\max[K - \bar{S}; 0]$; $\bar{S}$ is the average price of the underlying asset calculated over a specified time period. K is the strike price.	It is used by utilities that have to hedge average coal costs as the electricity tariffs they charge to customers are based on the average coal cost purchase price.
Barrier option	The option is activated or extinguished when the price of a reference asset (usually the underlying asset price) reaches a predetermined level, namely the barrier.	It is used in portfolio management to reduce the cost of hedging, and to allow the investor to readjust the hedge when circumstances change.
Spread option	For a call, $\max[P^1 - P^2 - K; 0]$; for a put, $\max[K - (P^1 - P^2); 0]$; P^1 and P^2 are the prices of commodities 1 and 2 at option expiration. K is the strike price.	It is used to hedge the risk of price change of a commodity from one location to another when the commodity is transferred. It is considered the spread between the commodity price in the two different markets.
Basket option	The underlying is a weighted sum or average of different assets that have been grouped together in a basket.	It is used when the price of a commodity depends on or adjusts according to an index.
Option to exchange one asset for another	The buyer has the option to choose the better of two commodities.	It is used to allow the holder to buy a commodity at prices related to another commodity.
Compound option	The underlying asset is an option.	It is used to lock in the premium of the option.
Digital option	The payoff depends on the occurrence of a predetermined event which is defined in terms of one or more prices.	Cash-or-nothing options, asset-or-nothing options.

Consider a number of fixing dates $T_1, ... T_n$, where at each date T_i one counterparty, who is the buyer of the swap, pays a fixed price K whereas the other counterparty, the seller, pays the variable price $S(T_i)$, usually given by a commodity index. The series of fixed payments represents the fixed leg of the swap; the flow of variable payments constitutes the floating leg. The discounted payoff at t of the fixed leg is calculated considering the forward prices $F(t, T_i)$ with delivery date $T_1, ..., T_i, ..., T_n$ as follows

$$\sum_{i=1}^{n} e^{-r(T_i - t)}(F(t, T_i) - k). \tag{1.2}$$

The fair swap price K is obtained by putting equal to zero equation (1.2) and solving for K:

$$K = \frac{\sum_{i=1}^{n} e^{-r(T_i - t)} F(t, T_i)}{\sum_{i=1}^{n} e^{-r(T_i - t)}}.$$

1.3.1 Plain Vanilla Swap

The plain vanilla swap is the basic type among swaps. It is traded over the counter and it has financial settlement. It is an agreement whereby a floating price is exchanged for a fixed price over a predetermined period. The floating price is the spot price of the underlying. The swap contract defines the volume, duration, fixed price of the underlying. The floating price is obtained at each exchange period from the market. Usually exchange periods are months, or quarters, or semesters, or years. There are fixed dates for payments between the inception and the maturity of the swap. The counterparties of the swap are two: the buyer who pays the fixed leg and receives the floating leg over the timelife of the swap, vice versa for the seller, see Figure 1.17. The swap involves counterparty risk, if it is not contemplated in the contract.

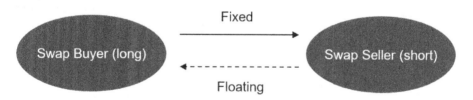

FIGURE 1.17: Swap structure

At every trading period, underlying price change impacts on the swap position and the following amount are considered:

Difference = Contracted monthly volume · (fixed price - floating price).

If "fixed price > floating price" then the buyer pays the difference.
If "fixed price < floating price" then the seller pays the difference.

Example 5 *The swap contract is usually used by the producer of electricity that wants to lock in the price of crude oil. See the example in Figure 1.18.*

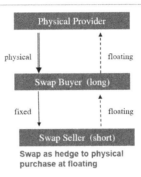

Brent Oil Swap: the fixed price is US$77

- Producers sell swaps to lock in their sales price
- Consumers of energy use swaps in order to stabilise the buying price

The difference at August 2018 is

50,000 bbl x (US$77.00-US$76.30)=US$35,000 ⟶ Who pays?

FIGURE 1.18: Example of swap on crude oil

We can say that at inception the swap has zero value and the commodity buyer buys a series of forward contracts plus an agreement to lend money at the implied forward rate. The swap contract market value changes over time because the forward price of the commodity and the interest rates will change over time. Even if forward prices and interest rates do not change, the value of the swap will remain zero only until the first swap payment is made. Once it is made, the market value of swaps can change over time due to the implicit borrowing and lending.

1.3.2   Other Swap Types

There are many types of commodity swaps; here we list the main ones.

- **Differential swap.** It is based on the difference between a fixed differential for two products, and the actual or floating differential over time. For example, this swap is used by refiners who want to lock in the margins of the refined products in a hedging perspective.

- **Participation swap.** The buyer who pays the fixed leg is 100% protected when prices rise above a predetermined price but "participates" in the downside. For example, it is used by a company that has to distribute gasoline and it sells the swap to benefit from any downside gasoline price movements.

- **The swaption.** It is a call or put option whose underlying asset is the swap on the index required.

- **Double-up swap.** It is a plain vanilla swap + call/put swaption usually sold by producers/consumers.

- **Extendable swap.** It is a double-up swap which gives the right to extend the swap, at the end of a predetermined period, for a further established period.

- **Complex swap.** The fixed price is a premium (or discount) built to fund (or realise) the purchase of options or to allow for the restructuring of a hedge portfolio.

- **Curve-lock and backwardation swaps.** They are based on the spread between different points on the commodity price curve. For example, it is used by a market participant who needs to lock into either backwardation or contango in the market.

- **Prepaid swap.** All the fixed leg payments are discounted at inception and paid to the swap seller.

1.4 Commodity Spreads

Commodity spreads are obtained by assuming both a long position and a short position in different contracts on the same and related commodities. An example is the intermarket spread. The intermarket spread involves the simultaneous purchase and sale of different but related commodities that have a reasonably stable relationship to each other. Opportunities for intermarket spread occur when commodities being traded are substitutes for each other or there are some other relationships that cause prices to be correlated. For example, random disturbances in supply and demand in cash and futures markets can cause futures prices to diverge and give rise to intermarket spread opportunities.

A classical example of intermarket spreads in commodity futures markets is the crack spread. The crack spread is the difference between the futures price of crude oil and an appropriate combination of futures prices of two petroleum products, that is heating oil and gasoline. The portfolio consists of a long position on three crude oil futures contracts and short positions on two gasoline futures contracts and one heating oil futures contract. It is a statistical arbitrage portfolio with the following time t value:

$$X_t = 3Z_t - 2v_t^1 - v_t^2, \quad t \geq 0, \tag{1.3}$$

X_t is the portfolio value, and Z_t, v_t^1 and v_t^2, respectively the futures prices of crude oil, gasoline and heating oil. The dynamics of X_t is shown in Figure 1.19. A crack spread position (buy crude oil and sell gasoline and heating oil) would be assumed when refined product prices are high relative to crude oil prices and are expected to fall. The refineries purchase crude oil and sell refined products in relatively fixed proportions. Therefore, the prices of crude oil, heating oil, and gasoline tend to move in a parallel fashion. When prices of refined products increase substantially above crude oil prices, it is convenient to purchase crude oil and sell refined products. This would cause the crack spread to narrow. On the contrary, when prices of refined

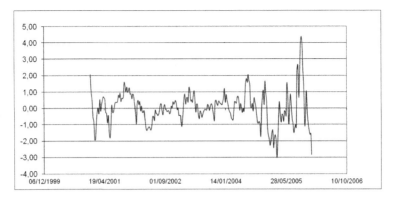

FIGURE 1.19: Crack spread dynamics

products fall below the crude oil price, the purchase of less crude oil and the running of the refinery at less than full capacity would lead to increase in the crack spread.

Other examples of statistical arbitrage portfolios are the spark spread and the frac spread. The first one mimics financially the generation costs of electricity for a specific facility and involves the simultaneous purchase of natural gas futures and the sale of electric futures. The second one is the difference between the price of gas liquids and natural gas.

1.5 Exercises

1. On January 15, the refiner X estimates it will need to purchase 10,000 barrels of crude oil on May 20. X decides to hedge price risk using a June NYMEX futures contracts. On January 15, the June futures price is USD 70.00/bbl. On May 20, the crude spot price is USD 73.10/bbl and the futures price is USD 71.5/bbl. On that day X can buy the required crude oil and closes out the futures contract. In this scenario, what is the effective price paid per barrel?

   a. USD 61.50

   b. USD 73.10

   c. USD 71.60

   d. USD 70.00

2. The spot price of gasoline is USD 0.60/gal. The cost of financing for the purchase of the contract is USD 0.20/gal per month; the cost of storage for the physical commodity is USD 0.15/gal per month. The price for a 1-month gasoline futures contract is USD 0.73/gal. In this scenario, what would be the best strategy to set up a profitable/ riskless spot/futures arbitrage trade?

    a. Buy spot gasoline at USD 0.60/gal and simultaneously sell an equal number of futures contracts units at USD 0.73/gal.

    b. Sell spot gasoline at USD 0.60/gal and simultaneously buy an equal number of futures contracts units at USD 0.73/gal.

    c. Buy spot gasoline at USD 0.60/gal, store it for one month and then sell it on the spot market.

    d. Do nothing; no strategy would result in a riskless profit.

3. On June 15, the spot price of crude oil is USD 97. The monthly storage cost id USD 0.60/bbl, and the annual risk-free interest rate is 2%. If the crude oil can be stored for six months but cannot be sold out of storage before the six-month storage term ends, what breakeven forward price per barrel supports the storage strategy?

    a. USD 100.3

    b. USD 97.97

    c. USD 101.57

    d. USD 95

4. Tom holds the following two option contracts on Brent futures:

    • Short one put option with strike price of USD 100.00;

    • Long one call option with strike price of USD 105.00.

If the options have the same maturity and the current Brent spot price is USD 103.50, what is the value of the combined positions?

    a. USD 2.5

    b. USD -2.5

    c. USD 0

    d. USD 5

5. Clark buys a monthly 100 MW on-peak power call option for a month that has 20 business days[4]. The strike price is USD 70/MWh and the premium is USD 5/MWh. Clark exercises the call option in a month when the average on-peak price is USD 80/MWh. What is the gross settlement amount for Clark?

    a. USD 320,000

    b. USD 160,000

    c. USD 240,000

    d. USD 480,000

---

[4] In a day there are 16 on-peak hours.

6. Which of the following statements currently explains a key characteristic of a bear spread option structure?

   a. The structure is created by a short call with strike price K_2 and a long call with strike price K_1, where $K_2 > K_1$.

   b. The maximum profit occurs when the underlying price equals the strike price of the call with the lower strike price.

   c. The maximum profit occurs when the underlying price equals the strike price of the call with the higher strike price.

   d. Upside profit potential is unlimited.

7. Bird Airways would use commodity swaps to protect against rising jet fuel prices. Which of the following statement(s) about the use of commodity swaps is/are correct?

   (a) Bird Airways will sell a commodity swap to hedge the risk of rising jet fuel prices.

   (b) Bird Airways will have counterparty credit exposure on commodity swap

   a. Both statements

   b. Neither statement

   c. Statement (a) only

   d. Statement (b) only

1.6 Answers

1. c.

2. d.

3. c.

4. a.

5. a.

6. b.

7. d.

Chapter 2

Spot Price Modelling

2.1 One-factor Models

2.1.1 Geometric Brownian Motion

The first model which could be used for modelling spot prices is the geometric Brownian motion (GMB). It is the classical framework applied in stock markets.

The spot price dynamics $S(t)$ is described by the following stochastic differential equation

$$\frac{dS(t)}{S(t)} = \mu dt + \sigma dW(t) \tag{2.1}$$

with initial value S_0. In equation (2.1), μ is the constant drift or expected return, σ is the constant volatility of the return and $W(t)$ is the Brownian motion. Under the risk-neutral probability measure, the drift rate is substituted by the risk-free interest rate r and the Brownian motion after changing numeraire is $\widetilde{W}(t)$. Applying Ito's lemma[1] we obtain a closed form formula for the spot price:

$$S(t) = S_0 e^{\left[\left(r - \frac{1}{2}\sigma^2\right)t + \sigma \widetilde{W}(t)\right]}. \tag{2.2}$$

The log return evolves according to the following stochastic differential equation:

$$d\log\frac{S(t)}{S_0} = \left[\left(r - \frac{1}{2}\sigma^2\right)dt + \sigma d\widetilde{W}(t)\right],$$

---

[1]See Itô [45].

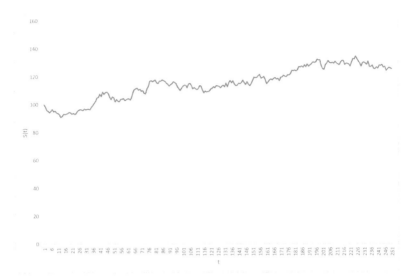

FIGURE 2.1: Geometric Brownian motion dynamics

and it is normally distributed as follows:

$$d\log S(t) \approx N\left(\left(r-\frac{1}{2}\sigma^2\right)dt, \sigma\sqrt{dt}\right). \quad (2.3)$$

In order to estimate the parameters of model (2.1) let us consider equation (2.1) in discrete time and use the linear regression. Then, we re-write (2.1) as

$$\Delta\log S(t) = \mu + \varepsilon(t) \qquad \varepsilon(t) \sim N(0, \sigma^2),$$

where $N(0, \sigma^2)$ represents the normal distribution with mean 0 and variance σ^2.

The dynamics of the GBM is plotted in Figure 2.1. It is obtained by implementing the stochastic process (2.2) in the Excel environment, as shown in Figure 2.2. The procedure for implementing the GBM according to Figure 2.2 consists of the following steps:

1. Establish the time steps (days, weeks, months, etc.) as in Column D;

2. Generate the random number draws from an uniform distribution [0,1] in Column E by using the function RAND();

3. Simulate the spot price according to formula (2.2) (Note that we have used μ instead of r).

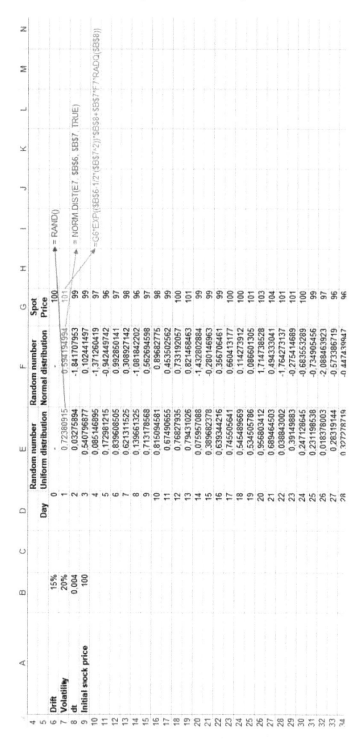

FIGURE 2.2: Excel implementation: Geometric Brownian motion

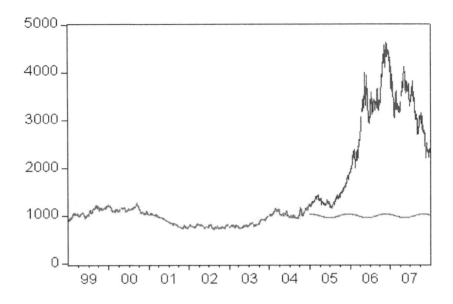

FIGURE 2.3: Zinc: Time series 4/01/1999-31/12/2007

Although the GBM is widely used also in commodity markets, for example for modelling the underlying financial options, it has huge drawbacks when it is applied in commodity markets. In fact, this classical model does not capture many features of several commodities, such as the mean-reverting effects, seasonality, spiky behaviour, stochastic volatility, and so on. In Figure 2.3 the dark grey line represents the actual zinc price, whereas the light grey line represents the price estimated by the GBM. The extreme event in 2004 is not captured by the classical model.

The one-factor model which could be seen as the extension of the GBM is the one proposed by Black [9]. This model can be easily adapted to commodities. The risk-neutral stochastic process for the spot price $S(t)$ is:

$$dS(t) = (r - \delta)S(t)dt + \sigma S(t)d\widetilde{W}(t), \qquad (2.4)$$

where r is the short-term interest rate, δ is the convenience yield and σ is the volatility of proportional changes in the spot energy prices.

2.1.2   Mean-reverting Process

A feature common to many commodity prices is the mean reversion. A price has a mean-reverting behaviour if it incorporates the tendency to gravitate towards a "normal" equilibrium price level that is usually governed by the cost of production and level of demand. Figure 2.4 shows the mean-reverting dynamics of the lead price.

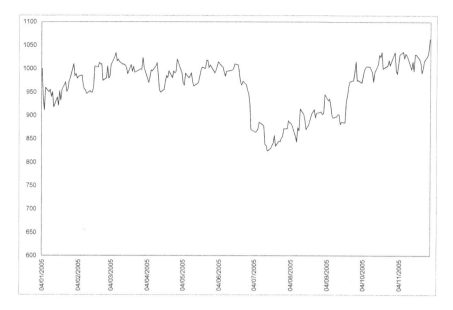

FIGURE 2.4: Lead price mean-reverting dynamics

Schwartz [60] proposes the following model for mean-reverting spot prices $S(t)$:

$$dS(t) = \alpha(\mu - \log S(t))S(t)dt + \sigma S(t)dW(t), \qquad S(0) = s, \qquad (2.5)$$

where s is the initial value, α is the speed of mean reversion, μ is the long-run mean level of log spot price, i.e., the price reverts to (e^μ), and σ is the diffusion volatility coefficient and $W(t)$ is the Brownian motion. A key property of a mean-reverting process is the halflife $\bar{t} = \frac{\log(2)}{\alpha}$. This is the time taken for the price to revert half way to its long-term level from its current level.

By defining $x(t) = \log S(t)$ and applying Ito's lemma we have the stochastic differential equation of the log price:

$$dx(t) = \alpha(\widehat{\mu} - x(t))dt + \sigma dWt, \qquad (2.6)$$

where $\widehat{\mu} = \mu - \frac{\sigma^2}{2\alpha}$.

Equation (2.6) in discrete-time is

$$\Delta x(t) = \alpha(\widehat{\mu} - x(t))\Delta t + \sigma\sqrt{\Delta t}\varepsilon(t), \qquad (2.7)$$

where $\varepsilon(t)$ is a normally distributed random number. Under the risk-neutral probability measure, $\widehat{\mu}$ changes to $\widehat{\mu} - \lambda$, where λ is the market price of risk.

The discrete-time mean-reverting model (2.7) is equivalent to the following simple linear model whose parameters can be estimated by using the ordinary least squares (OLS) regression:

$$\Delta x(t) = \alpha_0 + \alpha_1 x(t) + \sigma \varepsilon(t), \qquad (2.8)$$

where $\alpha_0 = \alpha \widehat{\mu} \Delta t$, $\alpha_1 = \alpha \Delta t$ and $\varepsilon(t) \sim N(0, \Delta t)$. The results of the linear regression applied to lead data plotted in Figure 2.4 are shown in Figure 2.5.

	delta_t	0,003968	=1/252
	alpha_0	0,279004	
	alpha_1	-0,040586	

		Annualised		
speed of mean-reversion	alpha	10,227595	-------------> 3,52	weeks half-life of mean reversion i.e. (log2/alpha)*52
long-run mean	mu	6,874432	-------------> 967,23	$ per tonns, i.e. exp(mu)
volatility coefficient	sigma	0,277694	=Standard Error*RADQ(52)	

REGRESSION RESULTS

SUMMARY OUTPUT

Regression Statistics	
Multiple R	0,133392307
R Square	0,017793508
Adjusted R Square	0,013504396
Standard Error	0,017493087
Observations	231

ANOVA

	df	SS	MS	F	Significance F
Regression	1	0,001269484	0,001269484	4,14853	0,042821187
Residual	229	0,070075852	0,000306008		
Total	230	0,071345336			

	Coefficients	Standard Error	t Stat	P-value	Lower 95%	Upper 95%	Lower 95.0%	Upper 95.0%
Intercept	0,279003583	0,136854723	2,038684361	0,04263	0,009348153	0,548659012	0,009348153	0,548659012
X Variable 1	-0,040585693	0,019926262	-2,036794083	0,042821	-0,079847946	-0,00132344	-0,079847946	-0,00132344

FIGURE 2.5: Results of OLS regression applied to equation (2.8)

2.2 Two-factor and Three-factor Models

Gibson and Schwartz [39] propose a two-factor model that has been used by many researchers and practitioners for modelling commodity spot prices. In particular, in the Gibson and Schwartz [39] model the first factor is the spot price $S(t)$ which is assumed to follow a geometric Brownian motion; the second factor is the instantaneous convenience yield $\delta(t)$ of the spot energy that follows a mean-reverting process. Assuming a given initial value for the spot price and the convenience yield we have:

$$
\begin{aligned}
dS(t) &= \mu_S S(t) dt + \sigma_S S(t) dW^1(t) \\
d\delta(t) &= \alpha(k - \delta(t)) dt + \sigma_\delta dW^2(t) \\
dW(t)^1 dW(t)^2 &= \rho_{S\delta} dt
\end{aligned}
$$

where

- μ_S is the drift of the spot price,

- σ_S is the instantaneous volatility of the spot price return,

- k is the long-term mean of the convenience yield,

- α is the speed of convergence of the convenience yield towards k,

- σ_δ is the instantaneous volatility of the convenience yield and $\rho_{S\delta}$ is the correlation between the two Brownian motions $dW(t)^1$ and $dW(t)^2$ associated to $S(t)$ and $\delta(t)$.

Under the risk-neutral probability measure, the drift terms become $\mu_S \to r$, the risk-free rate, and $\alpha(k - \delta(t)) \to \alpha(k - \delta(t)) - \lambda$, where λ is the market price of risk.

On the contrary, Schwartz [60] proposes the well-known model by assuming the following stochastic processes:

$$
\begin{aligned}
dS(t) &= (\mu - \delta(t))S(t)dt + \sigma_S S(t)dW^1(t) \\
d\delta(t) &= \alpha(k - \delta(t))dt + \sigma_\delta dW^2(t) \\
dW^1(t)dW^2(t) &= \rho_{S\delta}dt
\end{aligned}
\tag{2.9}
$$

where

- $S(t)$ is the spot price at time t,

- $\delta(t)$ is the time-t convenience yield,

- $\mu - \delta(t)$ is the drift of the spot price,

- σ_S is the instantaneous volatility of the spot price return,

- k is the long-term mean of the convenience yield,

- α is the speed of convergence of the convenience yield towards k,

- σ_δ is the instantaneous volatility of the convenience yield and $\rho_{S\delta}$ is the correlation between the two Brownian motions $dW^1(t)$ and $dW^2(t)$ associated to $S(t)$ and $\delta(t)$.

Under the risk-neutral probability measure, the drift terms become $\mu_S \to r$, the risk-free rate, and $\alpha(k - \delta(t)) \to \alpha(k - \delta(t)) - \lambda$, where λ is the market price of risk.

Another frequently used model is the one proposed by Schwartz and Smith [59]. Its characteristics are:

- Mean reversion in short-term prices;

- Uncertainty in the equilibrium level to which prices revert;

- Note: The Schwartz and Smith [59] model is different from the Schwartz [60] model because it does not explicitly model convenience yields. However, the state variables of the model can be represented in terms of a stochastic convenience yield.

Commodity spot price at time t can be decomposed into two stochastic factors as:

$$\log S(t) = \chi(t) + \xi(t), \tag{2.10}$$

where

- $\chi(t)$ is the short-term deviation in prices,

- $\xi(t)$ is the equilibrium price level.

The short-run deviation $\chi(t)$ is assumed to revert to zero following an Ornstein-Uhlenbeck process

$$d\chi(t) = -k\chi(t)dt + \sigma_\chi dW_\chi(t).$$

The equilibrium level $\xi(t)$ is assumed to follow a Brownian motion process

$$d\xi(t) = \mu_\xi dt + \sigma_\xi dW_\xi(t),$$

where $dW_\chi(t)$ and $dW_\xi(t)$ are correlated increments of standard Brownian motions, $dW_\chi dW_\xi = \rho_{\chi,\xi}$, k is the coefficient of mean reversion, μ_ξ is the drift of the process ξ and σ_ξ and σ_χ are the volatilities of the two factors. We can obtain the risk-neutral dynamics by substituting the drifts as follows: $-k\chi(t) \rightarrow -k\chi(t) - \lambda_\chi$ and $\mu_\xi \rightarrow -k\chi(t) - \lambda_\xi$, where λ_χ and λ_ξ are new parameters.

Schwartz [60] also proposes a three-factor model, which has spread among researchers and practitioners. The first factor is the spot price, the second factor is the instantaneous convenience yield of the spot energy and the third factor is the short-term rate. Under the risk-neutral probability measure their dynamics are described by the following stochastic processes:

$$
\begin{aligned}
dS(t) &= (r(t) - \delta(t))S(t)dt + \sigma_S S(t)d\widetilde{W}^1(t) \\
d\delta(t) &= \alpha(\overline{\delta} - \delta(t))dt + \sigma_\delta d\widetilde{W}^2(t) \\
dr(t) &= \beta(\overline{r} - r(t))dt + \sigma_r d\widetilde{W}^3(t),
\end{aligned}
$$

where

$$d\widetilde{W}^1(t)d\widetilde{W}^2(t) = \rho_{S\delta}dt \quad d\widetilde{W}^1(t)d\widetilde{W}^3(t) = \rho_{Sr}dt \quad d\widetilde{W}^2(t)d\widetilde{W}^3(t) = \rho_{r\delta}dt.$$

The volatility term structure depends on the volatility and speed of mean reversion of the three random variables and their correlations. Sometimes the interest rate is modelled in the Heath et al. [41] framework.

2.3 Jump-diffusion Models

Jumps in the spot prices have the following characteristics:

* Jumps are sudden, unexpected and discontinuous changes in energy prices;

* Jumps are a common feature of energy markets;
  In particular, power markets reflecting the nonstorability of the energy;

* Processes with jumps are characterised by excess kurtosis and non-zero skewness in the returns distribution, so non-Gaussian modelling solutions are required!

Figure 2.6 shows the behaviour of the electricity spot price that exhibits sudden jumps. The QQ test (see Section 6.3.1) can be used in order to identify the jump behaviour. Let us look at Figure 2.7, where we show the QQ test applied to the zinc spot returns. It can be assumed that the variable spot return follows two normal distributions:

* One for the central part and one for the tails;

* We have a mix of two normal distributions.

Merton [55] proposes the following dynamic model for the spot price $S(t)$:

$$dS(t) = \mu S(t)dt + \sigma S(t)dW^1(t) + kS(t)dZ(t),$$

where $Z(t)$ is the jump process, a discrete time process, i.e. Poisson process, μ the drift of the process, σ the volatility. The annualized frequency of jumps is given by

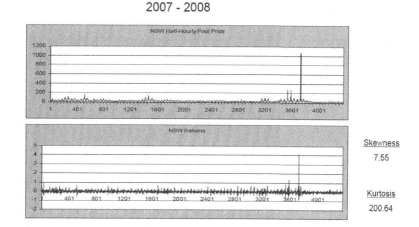

FIGURE 2.6: Australian New South Wales power prices

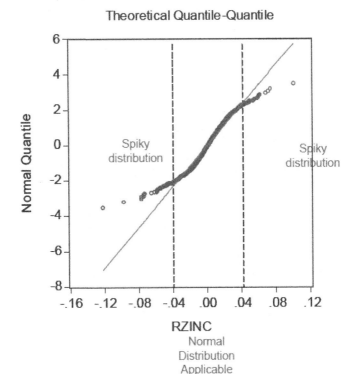

FIGURE 2.7: QQ test for zinc spot returns

ϕ, the average of jumps per year. The jump return size is k which is determined by natural logarithm of the jump returns being normally distributed $\ln(1+k) \sim N(\ln(1 + \bar{k}) - 1/2\gamma^2, \gamma^2)$, where γ is the standard deviation.

In order to simulate the model, we discretise the model in terms of spot price natural logarithm $x(t) = \log S(t)$ under the risk-neutral probability measure:

$$\Delta x(t) = (r - \phi\bar{k} - \frac{1}{2}\sigma^2)\Delta t + \sigma\sqrt{\Delta t}\varepsilon_{1,t} + (\bar{k} + \gamma\varepsilon_{2,t})\mathbf{1}_{\{u(t)>\phi\Delta t\}}, \qquad (2.11)$$

where r is the risk-free rate, $\varepsilon_{1,t}$ and $\varepsilon_{2,t}$ are independent standard normal random variables, $u(t)$ is a uniform(0,1) random sample and $\mathbf{1}_{\{u(t)>\phi\Delta t\}}$ is an indicator function.

In order to estimate the parameters of model (2.11) it could be used as a jumps-recursive filter estimation which attempts to identify and characterise lower frequency, higher volatility jump returns. It works recursively on return series and the filter estimation steps are:

1. Calculate sample standard deviation;

2. Identify returns greater than a threshold, typically 3 standard deviations, that depends on market data;

3. Remove these jump returns from the sample;

4. And repeat until convergence in jump intensity, mean jump size and jump volatility.

The Excel implementation is shown in Section 5.1. Consequently, a simple mean reversion jump diffusion (MRJD) could be obtained as a jump-augmented Schwartz model defined as follows:

$$dx(t) = \alpha(\hat{\mu} - x(t))dt + \sigma dW(t) + k_J dZ(t),$$

where $P(dZ(t) = 1) = \lambda dt, P(dZ(t) = 0) = (1 - \lambda)dt, k_J \sim N(\mu_J, \sigma_J^2)$, λ is the jump intensity, μ_J is the mean jump intensity and σ_J is the jump size standard deviation.

The MRJD estimation usually occurs through the maximum likelihood estimation (MLE) which may be used to estimate the parameters of the model. Given the mixture of normal distributions, the log returns process has tractable density function. The density function associated with log returns is given by

$$p(\Delta x(t); x(t), \theta) = (1 - \lambda \Delta t)\phi(\Delta x(t); \alpha(\hat{\mu} - x(t))\Delta t, \sigma^2 \Delta t) +$$
$$(\lambda \Delta t)\phi(\Delta x(t); \alpha(\hat{\mu} - x(t))\Delta t + \mu_J, \sigma^2 \Delta t + \sigma_J^2).$$

The MLE involves the maximization of the following log-likelihood function

$$L_T(\Delta x_0, ..., \Delta x_T; \theta) = \sum_{i=1}^{T} \ln[p(\Delta x_i; x_i, \theta)].$$

However the MLE has some limitations.

- MLE may not be able to sufficiently differentiate between diffusive and jump returns within the sample data; that is, given the mixture of normals, MLE could potentially consider a jump observation of one distribution to be a non-jump observation in the other.

- Additional topic for MRJD is joint-estimation of long-term mean reversion and short-term jump effects.

- MLE, although powerful, may still not perform an optimal estimation:

  - There exists some literature on alternative jump-diffusion estimation and the issues involved;
  - However, this literature has been extended with the development of alternative model specifications.

2.4 Seasonality Modelling

Many commodity markets contain a strong seasonal component either at the price level, or in the volatility. In particular, seasonality is a feature of many energy markets such as gas and power markets. Here there are seasonal highs and lows in energy demand that cause either winter-winter periodicity, or summer-summer periodicity, or winter-summer-winter periodicity. An example of a seasonal pattern is given in Figure 2.8.

The most common form of seasonality modelling is a periodic function composed of harmonically related trigonometric functions, combined by a weighted summation, so that the seasonal component of the spot price, $Y(t)$, is of type:

$$Y(t) = \alpha_0 + \sum_{i=1}^{N} [\alpha_i \cos(2\pi it) + \beta_i \sin(2\pi it)], \qquad (2.12)$$

where $i = 1$ represents annual periodicity in the cosine or sine function, $i = 2$ the semi-annual periodicity and $i = 3$ the quarterly periodicity in functions. Function (2.12) represents the smoothed trend line in spot trajectory in Figure 2.8. Parameters of function (2.12) are estimated through the ordinary least squares regression as:

$$Y(t) = \alpha_0 + \sum_{i=1}^{N} [\alpha_i \cos(2\pi it) + \beta_i \sin(2\pi it)] + \varepsilon(t), \quad \varepsilon(t) \sim N(0, \sigma^2), \qquad (2.13)$$

where regressors are non-stochastic and given by the cosine and sine values, and $N(0, \sigma^2)$ is the normal cumulative distribution with zero mean and variance σ^2, estimated on the residuals of regression. In Figure 2.9 we show the Excel estimation and implementation of function (2.12) on lead price data. The procedure follows the steps:

1. Obtain the fraction of the trade day by dividing the day (Column B) by 252, the days of a commercial year;

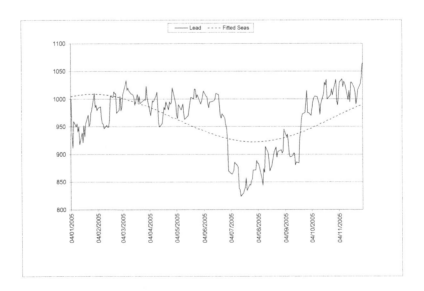

FIGURE 2.8: Seasonality in the spot price of lead

2. Obtain cosine and sine values with the fraction of the day in Columns D and E as inputs;

3. Apply the linear regression using as output the lead price (Column D) and as inputs the cosine and sine values (Columns D and E).

When we incorporate seasonal patterns in the volatility, we could use: *(i)* a function of time which repeats each seasonal cycle, so that the volatility follows a predictable seasonal pattern; *(ii)* a random variable with time varying parameters; *(iii)* a stochastic volatility.

Otherwise, when we use a mean reversion process for the spot price, the seasonal patterns could be described by the mean reversion rate. This practise is used for modelling electricity price. See, for example, Schmeck [58].

Finally, when we use a jump process for prices, the jump frequency and/or the jump volatility can be seasonal, as when the price takes into account the weather influence.

A seasonal model for spot prices is proposed by Pilipovic [56] and it is described in Figure 2.10.

2.5 Stochastic Volatility Model

Many commodity prices exhibit a certain variability in the volatility. Therefore, in order to have a more accurate pricing model, many authors have chosen to introduce a stochastic volatility in the model. We can cite the price modelling of crude oil, natural gas and agricultural commodities in some recent literature among others: Chen et al. [16], Chen and Xu [17], Hailemariam and Smyth [40], Li [50], Arismendi et al. [1], Benth [2], Trolle and Schwartz [62].

We recall Derman and Kani [24] who use the following risk-neutral spot price dynamics for describing the behaviour of an option underlying:

$$dS(t) = \mu(t)S(t) + \sigma(t, S(t))d\widetilde{W}(t), \qquad S_0 = s,$$

where $S(t)$ is the spot price, s the initial spot value, $\mu(t)$ is the drift depending only on time and $\sigma(t, S(t))$ is the local volatility function that is dependent on both spot price and time. A simple example of volatility function could be $\sigma(t, S(t)) = \beta_0 + \beta_1 S(t)^{\beta_2}$, where $= \beta_0$, β_1 and $\beta_2 > 1$ are constant.

Another useful model is the one proposed by Eydeland and Geman [27]. The spot price return variance is $V(t) = \sigma^2$ and it is stochastic, so that the spot price evolves according to the following model:

$$dS(t) = \mu S(t) + \sqrt{V(t)}S(t)dW^1(t) \qquad (2.14)$$

$$dV(t) = a(\overline{V} - V(t))dt + \xi\sqrt{V(t)}dW^2(t),$$

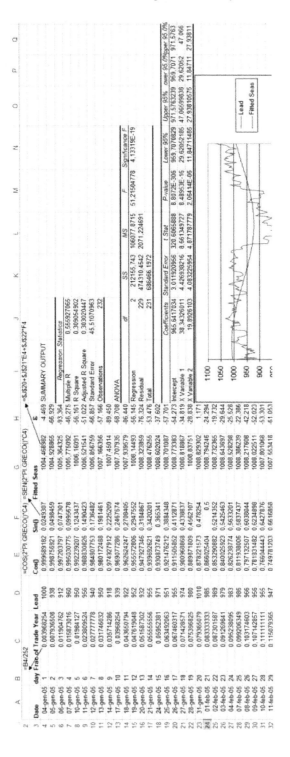

FIGURE 2.9: Excel implementation: Seasonal spot component estimation through formula through equation (2.13)

Assume spot price is a function of the underlying spot price plus seasonal factors:

$$S_t = S_t^{Und} + \sum_{n=1}^{3} \beta_n e^{-\gamma_n (rfc(t - t_n^C))^2}$$

where

S_t = spot price at time t

S_t^{Und} = underlying spot price value

β_n = annual seasonality magnitude

t_n^C = annual seasonality centering parameter (time of the annual peak)

γ_n = seasonal decay parameter

rfc = an annually repetitive function; it returns the annualized time to or from the closest annual center, t_n^C, for that particular factor.

Seasonal factors:
1. Summer
2. Winter
3. Factor that captures any additional repetitive annual event behavior (i.e. additional peaks in summer or winter, hump in the fall)

FIGURE 2.10: Seasonal model for spot price by Pilipovic [56]

where $\overline{V}$ is the long-term level, a is the speed of mean reversion, $\xi \sqrt{V(t)}$ is the absolute volatility of the variance that is proportional to the square root of the variance and $\rho = corr(dW^1(t), dW^2(t))$.

Discretising the time in the model (2.14) expressed in terms of the spot price natural logarithm $x(t) = \log S(t)$, we obtain

$$\Delta x(t) = (\mu - \frac{1}{2}V(t)^2)\Delta t + \sqrt{V(t)}\sqrt{\Delta t}\varepsilon_{1,t}, \qquad (2.15)$$

$$\Delta V(t) = a(\overline{V} - V(t))\Delta t + \xi \sqrt{V(t)\Delta t}\varepsilon_{2,t},$$

where $\varepsilon_{1,t}$ and $\varepsilon_{2,t}$ are independent standard normal random variables with $\rho = corr(\varepsilon_{1,t}, \varepsilon_{2,t})$. We can then simulate the model taking into consideration that $\omega_{1,t}$ and $\omega_{2,t}$ are two independent draws from a standard normal distribution and ρ is the correlation coefficient between the two variables. If we assume a bivariate distribution, we have: $\varepsilon_{1,t} = \omega_{1,t}$ and $\varepsilon_{2,t} = \rho \omega_{1,t} + \sqrt{1 - \rho}\omega_{2,t}$. In Figure 2.11 we implement the two correlated standard normal random variables $\varepsilon_{1,t}$ and $\varepsilon_{2,t}$, according to the following steps:

1. Calculate the logarithm of the price in Column C;

2. Calculate the moving standard deviation in Column D;

3. Calculate the deviations of the standard deviation in Column E;

4. Determine the correlation between the returns and the delta standard deviations;

5. Generate two draws from a standard normal distribution for each day, Columns G and H;

6. Obtain the two correlated draws, Columns I and J;

7. Apply formula (2.15).

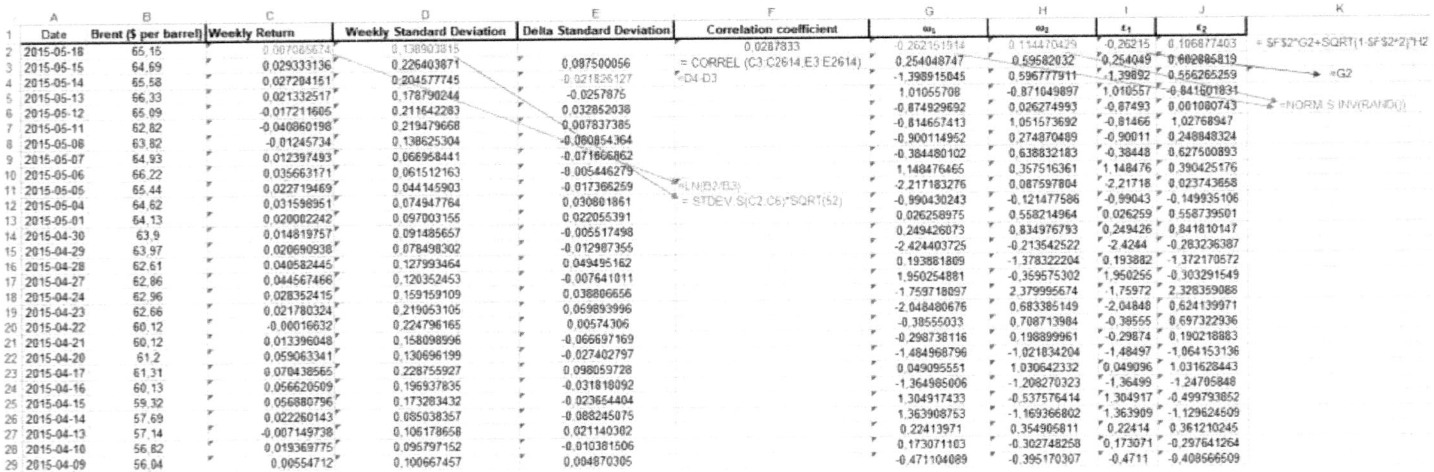

	A	B	C	D	E	F	G	H	I	J	K
1	Date	Brent ($ per barrel)	Weekly Return	Weekly Standard Deviation	Delta Standard Deviation	Correlation coefficient	ω_1	ω_2	ε_1	ε_2	
2	2015-05-18	65,15	0,007085674	0,138903815		0,0287833	-0,262151514	0,114470429	-0,26215	0,106877403	= $F$2*G2+SQRT(1-$F$2^2)*H2
3	2015-05-15	64,69	0,029333136	0,226403871	0,087500056	= CORREL (C3:C2614,E3:E2614)	0,254048747	0,59582032	0,254049	0,602885819	
4	2015-05-14	65,58	0,027204151	0,204577745	-0,021826127	=D4-D3	-1,398915045	0,596777911	-1,39892	0,556265259	=G2
5	2015-05-13	66,33	0,021332517	0,178790244	-0,0257875		1,01055708	-0,871049897	1,010557	-0,841601831	
6	2015-05-12	65,09	-0,017211605	0,211642283	0,032852038		-0,874929692	0,026274993	-0,87493	0,001080743	=NORM.S.INV(RAND())
7	2015-05-11	62,82	-0,040860198	0,219479668	0,007837385		-0,814667413	1,051573692	-0,81466	1,02768947	
8	2015-05-08	63,82	-0,01245734	0,138625304	-0,080854364		-0,900114952	0,274870489	-0,90011	0,248848324	
9	2015-05-07	64,93	0,012397493	0,066958441	-0,071666862		-0,384480102	0,638832183	-0,38448	0,627500893	
10	2015-05-06	66,22	0,035663171	0,061512163	-0,005446279		1,148476465	0,357516361	1,148476	0,390425176	
11	2015-05-05	65,44	0,022719469	0,044145903	-0,017366259	=LN(B2/B3)	-2,217183276	0,087597804	-2,21718	0,023743658	
12	2015-05-04	64,62	0,031598951	0,074947764	0,030801861	= STDEV.S(C2:C6)*SQRT(52)	-0,990430243	-0,121477586	-0,99043	-0,149935106	
13	2015-05-01	64,13	0,020002242	0,097003155	0,022055391		0,026258975	0,558214964	0,026259	0,558739501	
14	2015-04-30	63,9	0,014819757	0,091485657	-0,005517498		0,249426073	0,834976793	0,249426	0,841810147	
15	2015-04-29	63,97	0,020690938	0,078498302	-0,012987355		-2,424403725	-0,213542522	-2,4244	-0,283236387	
16	2015-04-28	62,61	0,040582445	0,127993464	0,049495162		0,193881809	-1,378322204	0,193882	-1,372170572	
17	2015-04-27	62,86	0,044567466	0,120352453	-0,007641011		1,950254881	-0,359575302	1,950255	-0,303291549	
18	2015-04-24	62,96	0,028352415	0,159159109	0,038806656		-1,759718097	2,379995674	-1,75972	2,328359088	
19	2015-04-23	62,66	0,021780324	0,219053105	0,059893996		-2,048480676	0,683385149	-2,04848	0,624139971	
20	2015-04-22	60,12	-0,00016632	0,224796165	0,00574306		-0,38555033	0,708713984	-0,38555	0,697322936	
21	2015-04-21	60,12	0,013396048	0,158098996	-0,066697169		-0,298738116	0,198899961	-0,29874	0,190218883	
22	2015-04-20	61,2	0,059063341	0,130696199	-0,027402797		-1,484968796	-1,021834204	-1,48497	-1,064153136	
23	2015-04-17	61,31	0,070438565	0,228755927	0,098059728		0,049095551	1,030642332	0,049096	1,031628443	
24	2015-04-16	60,13	0,056620509	0,196937835	-0,031818092		-1,364985006	-1,208270323	-1,36499	-1,24705848	
25	2015-04-15	59,32	0,056880796	0,173283432	-0,023654404		1,304917433	-0,537576414	1,304917	-0,499793852	
26	2015-04-14	57,69	0,022260143	0,085038357	-0,088245075		1,363908753	-1,169366802	1,363909	-1,129624509	
27	2015-04-13	57,14	-0,007149738	0,106178658	0,021140302		0,22413971	0,354905811	0,22414	0,361210245	
28	2015-04-10	56,82	0,019369775	0,095797152	-0,010381506		0,173071103	-0,302748258	0,173071	-0,297641264	
29	2015-04-09	56,04	0,00554712	0,100667457	0,004870305		-0,471104089	-0,395170307	-0,4711	-0,408566509	

FIGURE 2.11: Excel implementation: Stochastic volatility model

2.6 Regime-switching Models

In the literature, regime-switching models are used to model electricity spot price evolution, (see the first models of Mari [53], Huisman and Mahieu [44]) such that the normal stable motion is distinguished from the spike regime and switches between regimes are controlled by a transition probability measure. Mari [53] models the log spot price (of 1 MWh of electricity) dynamics by the sum of two components:

$$s(t) = f(t) + x(t),$$

where $f(t)$ is a function that describes the price seasonality (daily, weekly, monthly, yearly), whereas $x(t)$ is a random component reflecting unpredictable movements of the prices. In the two-regime approach, one regime drives the stable motion during normal periods and the second regime is used to account for turbulent periods with high volatility, high values of mean- reversion rate, jumps and short-lived spikes. Therefore, the first regime is characterised by a mean-reverting diffusion process (a), whereas the second one is represented by mean-reverting jump-diffusion dynamics (b), as follows:

$$dx(t) = \begin{cases} -\alpha_0 x(t)dt + \sigma_0 dW_0(t), & \text{(a)} \\ (\mu - \alpha_1 x(t))dt + \sigma_1 dW_0(t) + JdQ(t), & \text{(b)} \end{cases}$$

where $q(t)$ is a Poisson process with constant intensity λ. The random jump amplitude J is distributed according to a normal random variable, $J \sim N(\mu_J, \sigma_J)$, with mean μ_J and standard deviation σ_J. The Brownian motion W_0 and the Poisson process Q are independent and independent of the jump amplitude. The switches between regimes in both models are controlled by the following one-period transition probabilities matrix:

$$\pi = \begin{pmatrix} 1 - \gamma dt & \eta dt \\ \gamma dt & 1 - \eta dt \end{pmatrix},$$

where γdt denotes the transition probability for the switching from the base state to the second regime in the infinitesimal time interval $[t, t + dt]$ and ηdt is the probability for the opposite transition. γ and η are assumed to be constant.

The model will be extended by introducing three regimes. We recall among others some papers in regime-switching literature: Weron et al. [64], Janczura and Weron [46], Chen and Insley [18] and Chevallier et al. [19].

2.7 Exercises

1. When commodity spot return distribution exhibits excess kurtosis and non-zero skewness, the model used for spot prices is:

   (a) Jump-diffusion model;
   (b) Mean-reverting model;

    (c) Geometric Brownian motion;

    (d) Brownian motion.

2. Which of the following statements is true:

    (a) Many commodity markets contain a strong seasonal component either at the price level, or in volatility.

    (b) In commodity markets the seasonal component is always contained in volatility.

    (c) In commodity markets the seasonal component is always contained in price level.

    (d) None of the previous statements is true.

3. What is the halftime of a mean-reverting process?

    (a) It is the time taken for the price to revert to zero.

    (b) It is the time taken for the price to revert half way to its long-term level from zero.

    (c) It is the time taken for the price to revert half way to zero from its current level.

    (d) It is the time taken for the price to revert half way to its long-term level from its current level.

4. In the two factor model of Schwartz [60] as the time to maturity approaches large values, the volatility of futures converges to:

    (a) zero.

    (b) infinity.

    (c) a fixed value depending on model paramenters.

    (d) the spot volatility.

2.8 Answers

  1. (a)

  2. (a)

  3. (d)

  4. (c)

Chapter 3

Forward Price Modelling

3.1   Forward/Futures Valuation

Commodity pricing models are mostly calibrated using panel data of futures prices. To price derivatives, as futures, only the risk-neutral or risk-adjusted distribution of the future spot prices is required. Futures prices are the expected spot prices under the risk-neutral distribution. Therefore, the time-t price of a futures/forward contract with maturity T is

$$F(t,T) = E_t^Q[S(T)], \tag{3.1}$$

where Q is the risk-neutral probability measure, and $E_t^Q[\cdot]$ is the expected value conditional on the information available up to time t and $S(T)$ is the spot price at maturity. However, as we have seen in Chapter 1, under the true distribution there is a relationship that links the spot price and forward price, namely:

$$F(t,T) = E_t[S(T)]e^{\lambda(T-t)}, \tag{3.2}$$

that is, the forward price and the expected spot price differ only on the risk premium λ.

Therefore, given the expected spot price, we can obtain the forward price through models discussed in Chapter 3 and by applying one of the two relationships (3.1) and (3.2).

Black [9] shows that given the spot process of equation (2.4) under appropriate boundary conditions, the futures/forward price is

$$F(t,T) = S(t)e^{(r-\delta)(T-t)},$$

where r is the interest rate and δ is the convenience yield. Furthermore, by applying Ito's lemma, we find that the volatilities of the futures prices are constant and equal to the volatility of the spot price: $\sigma(t,T) = \sigma$. The main drawback of the Black model is that the futures/forward price volatility is constant and equal to the one of the spot price. On the contrary, futures/forward price volatilities vary with maturity, typically declining relative to the spot price volatility for increasing maturities and often exhibiting seasonal patterns.

According to the model of Schwartz [60], given the spot price dynamics (2.9) under the appropriate boundary conditions, the futures/forward price is

$$F(t,T) = exp\left[e^{-\alpha(T-t)}\ln S(t) + (1 - e^{-\alpha(T-t)})\left(\mu - \lambda - \frac{\sigma^2}{2\alpha}\right) + \frac{\sigma^2}{4\alpha}(1 - e^{-2\alpha(T-t)})\right],$$
(3.3)

where α is the speed of mean reversion, μ is the long-term level, σ is the spot price volatility and λ is the market price of risk.

Applying Ito's lemma to equation (3.3), the model provides a futures volatility term structure, such that the futures volatilities are proportional to the volatility of the spot price: $\sigma(t,T) = \sigma e^{-\alpha(T-t)}$. This indicates that the volatility decreases as

- the speed of mean reversion increases,

- the maturity of the forward contract increases.

In the Schwartz and Smith [59] model the futures contract at time t can be computed using formula (3.1), and so the dynamics (2.10) of $S(t)$ under the risk-neutral measure is required. Schwartz and Smith [59] assume that the spot price $S(t) = \chi(t) + \xi(t)$ involves risk-neutral processes of the following form

$$d\chi(t) = -(k\chi(t) + \lambda_\chi)dt + \sigma_\chi d\widetilde{W}_\chi(t),$$
(3.4)

$$d\xi(t) = (\mu_\xi - \lambda_\xi)dt + \sigma_\xi d\widetilde{W}_\xi(t),$$
(3.5)

where

$$d\widetilde{W}_\chi d\widetilde{W}_\xi = \rho_{\chi,\xi}dt,$$

where $d\widetilde{W}_\chi(t)$ and $d\widetilde{W}_\xi(t)$ are correlated increments of standard Brownian motions. We recall that under the risk-neutral measure $\chi(t)$ reverts to $\frac{\lambda_\chi}{k}$, under the empirical measure $\chi(t)$ reverts to 0.

Then the expression for the forward curve at any time t in the risk-neutral world is

$$\log F(t,T) = e^{-k(T-t)}\chi(t) + \xi(t) + A(T-t),$$
(3.6)

where

$$A(T-t) = (\mu_\xi - \lambda_\xi)(T-t) - (1 - e^{-k(T-t)})\frac{\lambda_\chi}{k} +$$
$$\frac{1}{2}\left[(1 - e^{-2k(T-t)})\frac{\sigma_\chi^2}{2k} + \sigma_\xi^2(T-t) + (1 - e^{-k(T-t)})\frac{\rho_{\chi,\xi}\sigma_\chi\sigma_\xi}{k}\right],$$

where $\chi(t)$ is the short-term deviation in prices, $\xi(t)$ is the equilibrium price level, k is the coefficient of mean reversion, μ_ξ is the drift of the process ξ and σ_ξ and

σ_χ are the volatilities of the two factors. Therefore, the parameters of the spot price component dynamics in (3.4) and (3.5) are usually obtained by calibrating the model for the forward price (3.6) using futures panel data. Futures prices are the expected spot prices under the risk-neutral distribution, so the parameters under risk-neutral distribution are usually precisely estimated and model (3.6) is very successful in fitting observed futures term structures and its dynamics.

In Figure 3.1 we calibrate the parameters of the Schwartz and Smith [59] model using fuel oil futures data in the Excel environment. In particular, after calculating the futures log-prices in Columns D and E according to price data and formula (3.6), we obtain the squared errors and the pricing errors in Columns F and G. Then, we apply the Excel Solver to minimize the sum of the squared pricing errors.

The statistical method used to estimated the parameters in Figure 3.1 is the non-linear ordinary least squares. The implied estimation of spot parameters through the forward prices is done as follows:

$$\theta = \{\alpha, \mu, \lambda, \sigma\}$$

$$\min_\theta \frac{1}{N} \sum_{i=1}^{N} [F^{mod}(t, T_i, \theta) - F^{mkt}(t, T_i)]^2; \quad N \geq 4,$$

where $F^{mod}(\cdot)$ and $F^{mkt}(\cdot)$ are respectively the price obtained from the model implementation and the market price. The front-end forward contract could be used as a proxy for the spot price!

Another approach for obtaining the forward price is given by the application of the cost-of-carry relationship. It is based on the no-arbitrage assumption and implies that the futures price of a commodity must be equal to the cost of acquiring the physical commodity and carrying it until the future maturity T:

$$F(t, T) = S(t) e^{[r(t) + m(t) - c(t)](T - t)},$$

where $r(t)$ is the riskless interest rate on the date t, $m(t)$ is the storage cost (per unit of time and per dollar's worth of commodity) and $c(t)$ is the convenience yield, all continuously compounded. If we define $\gamma(t) = c(t) - m(t)$ we obtain

$$F(t, T) = S(t) e^{[r(t) - \gamma(t)](T - t)}.$$

Among other models, we recall Lucia and Schwartz [52] and Sørensen [61] who propose stochastic factor models. We start by defining the process for the commodity spot price:

$$S(t) = s(t) + X(t) + L(t),$$

where $X(t)$ and $L(t)$ are respectively the short- and the long-term price components and $s(t)$ is a possible deterministic seasonal component of the spot price. The dynamics of these components should be specified. The futures price is then obtained by applying formula (3.1).

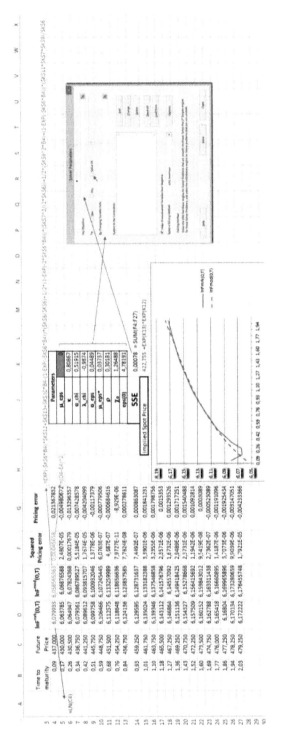

FIGURE 3.1: Excel implementation: Parameters calibration of the Schwartz and Smith [59] model

3.2 Forward Price Models

Forward curve dynamics in a risk-neutral world are described by the following stochastic differential equation

$$\frac{dF(t,T)}{F(t,T)} = \sigma(t,T)d\widetilde{W}(t),$$

where $\sigma(t,T)$ is the volatility function and $\widetilde{W}(t)$ is a Brownian motion. We can use a simple negative exponential volatility function for forward prices: $\sigma(t,T) = \sigma e^{-\alpha(T-t)}$ as in Figure 3.2.

One of the most used models for the forward curve for commodities (seasonal and non seasonal) is the multifactor model. In the literature it is widely used. We recall for example Clewlow and Strickland [21], Benth et al. [5], and Fanelli and Schmeck [31]. The futures prices are modelled directly under the risk-neutral probability measure. This approach is useful for derivative pricing, but not for trading strategies. The general model is similar to the Heath et al. [41] approach for modelling yield curve, that is:

$$\frac{dF(t,T)}{F(t,T)} = \sum_{i=1}^{n} \sigma_i(t,T;\cdot)d\widetilde{W}_i(t), \tag{3.7}$$

where n is the number of factor risks, $\widetilde{W}_i(t)$, $i = 1,...,n$ are Brownian motions representing sources of uncertainty and $\sigma_i(t,T;\cdot)$ are volatilities associated with risk factors. The point in the argument of the volatility function, $\sigma_i(t,T;\cdot)$, indicates that

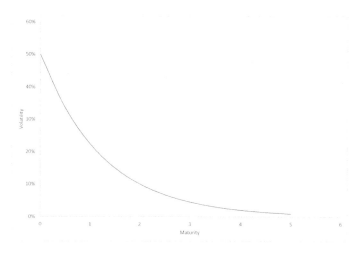

FIGURE 3.2: The simple negative exponential volatility function for forward prices

the volatility could depend on some factors, such as the price level, yield curve, and similar factors. By integrating equation (3.7), we have the forward curve at time t in terms of its initially observed state (time 0) and integrals of the volatility functions:

$$F(t,T) = F(0,T)exp\left[\sum_{i=1}^{n}\left(-\frac{1}{2}\int_{0}^{t}\sigma_i^2(u,T)du + \int_{0}^{t}\sigma_i(u,T)d\widetilde{W}_i(u)\right)\right].$$

By applying Ito's lemma we obtain the logarithmic process

$$d\log F(t,T) = -\frac{1}{2}\sum_{i=1}^{n}\sigma_i(t,T)^2 dt + \sum_{i=1}^{n}\sigma_i(t,T)d\widetilde{W}_i(t).$$

And we could discretize for small time changes Δt to have

$$\Delta \log F(t,t+\tau_j) = -\frac{1}{2}\sum_{i=1}^{n}\sigma_i(t,t+\tau_j)^2 \Delta t + \sum_{i=1}^{n}\sigma_i(t,t+\tau_j)\Delta\widetilde{W}_i(t).$$

Therefore, the natural logarithm of the forward prices with relative maturities τ_j, $j = 1,...,m$ are jointly normally distributed.

All the models proposed for spot prices could be easily adapted or used for forward pricing. We summarise all the pricing approaches seen in Chapter 2, in the following model. Under the equivalent martingale measure we could propose the following forward price dynamics in a regime-switching framework:

$$dF(t,T) = \begin{cases} F(t,T)\left[\alpha(t,T)dt + \sigma(t,T)d\widetilde{W}(t)\right] \\ F(t,T)\left[\alpha(t,T)dt + \sigma(t,T)d\widetilde{W}(t) + Jd\widetilde{P}(t)\right] \end{cases},$$

where $\alpha(t,T)$ is the drift function, $\sigma(t,T)$ is the volatility function, $\widetilde{P}(t)$ is a Poisson process with constant intensity and J is a stochastic jump amplitude distributed according to a normal random variable, $J \sim N(\mu_J, \sigma_J)$. $\widetilde{W}(t)$ is a Brownian motion. $\widetilde{W}(t)$ and $\widetilde{P}(t)$ are independent and are also independent of the jump amplitude. The switches between regimes are controlled by a one-period transition probability matrix:

$$\pi = \begin{pmatrix} 1-pdt & qdt \\ pdt & 1-qdt \end{pmatrix},$$

where pdt is the transition probability for the switching from the normal state to the spike state in the infinitesimal time interval $[t,t+dt]$ and qdt is the probability for the opposite transition. In the matrix, the diagonal terms give the probability of remaining in any given state during the interval $[t,t+dt]$; the off-diagonal terms represent the transition probability to the other state in the same time interval. We assume that p and q are constant.

We would like to give a deeper look at modelling the electricity. The electricity is a flow commodity, in the sense that it is useful for practical purposes only if it is delivered over a period of time, so forward contracts are written on the average of the hourly system price established on the day ahead market over a specified delivery period. During the delivery period the contract is settled in cash against the system price, so that it is more proper to refer to financial electricity contracts as swap contracts, exchanging a floating spot price against a fixed price. Furthermore, the "time to delivery" replaces the term "time to maturity" used for fixed maturity contracts. The main feature of electricity is that it has very limited storage possibility. From this all the features of the spot electricity price follow. Geman and Roncoroni [38] reveal the key features of the electricity price: mean reversion toward a level that represents marginal costs, seasonality, the existence of small random moves around the average trend, which represent the temporary supply/demand imbalances in the network; spiky behaviour. According to Benth et al. [5], the non-storability of the electricity implies the breakdown of the spot-forward relationship and the incompleteness of the market, making spot-forward hedging impossible. Furthermore, the existence of a delivery period puts restrictions on the class of spot models feasible for analytical pricing. Thus, instead of explaining the forward and swap price by the underlying spot, we model directly the forward/futures price adopting the Heath et al. [41] approach from interest rate theory. This approach suggests to directly assume a dynamics or the forward and swap price evolution either in terms of market dynamics or under the risk-neutral measure. A discussion of the Heath et al. [41] approach to general energy markets is in Clewlow and Strickland [21]. The authors as well as Bjerksund et al. [7] suggest to model forward contracts, while Benth and Koekebakker [3] consider the actual contracts traded in the market, electricity futures/swaps. Clearly the implementation of the Heath et al. [41] model is straightforward in the case of forward price modelling, but these prices are not directly observable in the market and they have to be estimated. Benth et al. [4] propose an algorithm for constructing a smooth curve of forward price from swap contracts in order to model the forward curve in the Heath et al. [41] framework. This approach makes data dependent from the algorithm chosen. Frestad [35, 34] argues there exists a correlation structure across a broader set of electricity contracts and he demonstrate that in the Nord Pool, electricity swap returns can be explained by common and unique risk factors. Consequently, forward price modelling should be preceded by a statistical analysis about the correlation among contracts with different delivery periods. An alternative method to model the forward or future curve is suggested by Benth and Koekebakker [3]. They propose a modelling approach for electricity swap prices of the Nord Pool market similar to the LIBOR approach in interest rate theory. Their idea is to construct a dynamics for the traded contracts matching the observed volatility term structure.

In electricity markets, the commodity is delivered over a period, so we obtain the electricity swap price, $F_s(t, \tau_1, \tau_2)$ for a contract that settles continuously during the delivery period $[\tau_1, \tau_2]$ integrating the forward curve over the delivery period

$$F_s(t, \tau_1, \tau_2) = \int_{\tau_1}^{\tau_2} \hat{w}(u, \tau_1, \tau_2) F(t, u) du, \tag{3.8}$$

where, $F(t,u)$ is the istantaneous forward price, and

$$\hat{w}(u,s,t) = \frac{w(u)}{\int_s^t w(v)dv},$$

is the weight function connecting forward with swaps and, assuming that r is the risk-free interest rate, we set $w(u) = e^{-ru}$ so that

$$\hat{w}(u,s,t) = \frac{re^{-ru}}{e^{-rs} - e^{-rt}}.$$

3.3   Modelling the Seasonality

Borovkova and Geman [11] propose a multifactor model for modelling seasonality:

- They assume that the forward curve contains liquid maturities up to one year (12 months) or an integer number of years;

- The geometric average of the observed forward price at date t is

$$\overline{F}(t) = \sqrt[N]{\prod_{T=1}^{N} F(t,T)},$$

where N is the most distant maturity ($N = 12 \cdot k, k = 1,2,...$). This average is a non-seasonal quantity, but a proxy of the overall level of futures market;

- The seasonal cost-of-carry relationship is

$$F(t,T) = \overline{F}(t)e^{[s(t)-\gamma(t)](T-t)},$$

where $s(T) = s(M)$, $M = 1,...,12$ is the seasonal premia, attached to calendar months and deterministic, it measures the seasonal effect associated with periods of high demand or low supply; $\gamma(t,T-t)$ is the stochastic premium (or stochastic cost-of-carry), indexed by the time to maturity $T-t$.

See Geman [37] for a generalization of the model.

Another way to take into consideration the seasonality is by representing the behaviour of the forward curve as follows:

$$\frac{dF(t,T)}{F(t,T)} = \sigma_S(t) \sum_{i=1}^{n} \sigma_i(t,T)d\widetilde{W}_i(t),$$

where $\sigma_S(t)$ denotes the spot price volatility at time t, $\sigma_i(t,T)$ denotes the n maturity dependent volatility functions and $\widetilde{W}_i(t)$ are Brownian motions under the risk-neutral measure. In this way, the maturity structure of the volatility functions is normalised by the spot volatility. To estimate this model:

- Estimate the spot volatility (i.e., using a rolling 30-day sample standard deviation);

- Divide the daily forward price returns by the daily spot volatility;

- Apply the Principal Component Analysis[1].

Furthermore, particular functions could be used in order to capture the seasonal behaviour of the forward price. We cite Fanelli et al. [32] who use a path-dependent volatility that captures the seasonality feature of futures rates. The volatility is the function

$$\sigma(t,T) := S(t,T) + X(t,T),$$ (3.9)

where $S(t,T)$ is the seasonal term and $X(t,T) = [f(t)]^u$, with $u \in \mathbf{R}$, is the path-dependent term. The seasonal term has the following form

$$S(t,T) := \sum_{i=1}^{12} (\sigma_2^i + \sigma_1^i e^{-\lambda_i(T-t)}) D_i,$$

where dummy variables D_i allow to estimate the coefficients σ_1^i and σ_2^i for every month of the year. Parameter σ_1 represents the short-term volatility coefficient, σ_2 the long-term volatility coefficient, and λ the time decay. Thereby, the volatility function can be adapted, through the coefficients, reflecting the monthly seasonality.

The seasonal term could have other different functional forms, such as, for example, the one proposed by Fanelli and Schmeck [31], namely:

$$S(t) = s_1 + s_2 \cos\left(\frac{s_3 + 2\pi t}{12}\right),$$ (3.10)

where s_1, s_2 and s_3 are constants to be estimated on data.

3.4 Exercises

1. Which of the following statements are true?

   Statement I Under the true distribution the forward price and the expected spot price differ only on the risk premium.

   Statement II The forward price can be obtained through spot price models.

   (a) Only Statement I;
   (b) Only Statement II;
   (c) Both;
   (d) Neither.

---

[1] See Hotelling [42] and Hotelling [43] for theory on the Principal Component Analysis.

2. In the Schwartz and Smith [59] model the forward curve at any time in the risk-neutral world depends on:

   (a) The short-term deviation in prices and the equilibrium price level;

   (b) The short-term deviation in prices and the convenience yield;

   (c) The short-term deviation in prices and the risk premium;

   (d) The short-term deviation in prices and the seasonality.

3. The Schwartz and Smith [59] model is calibrated by using:

   (a) The call option prices;

   (b) The spot prices;

   (c) The swap prices;

   (d) The futures prices.

4. What are the two components of the futures volatility function in Fanelli et al. [32] model?

   (a) A seasonal component and a path-dependent component;

   (b) A short-term component and a path-dependent component;

   (c) A seasonal component and a short-term component;

   (d) A constant component and a path-dependent component.

3.5   Answers

1. (c)

2. (a)

3. (d)

4. (a)

Chapter 4

Derivative Valuation

4.1 Introduction to Valuation Models

A financial derivative is a contract which is derived from an underlying asset. Its fair value is defined as the price for which a neutral market participant would be willing to buy or sell the contract. The contract is valuated to market prices on a regular basis in order to obtain a marked-to-market valuation. This valuation guarantees that the market is arbitrage free; that is, it is not possible to make a profit without taking any risk.

In order to define an arbitrage possibility, we start from the definition of self-financing strategy. A self-financing strategy is a portfolio (combination of assets) from 0 to T which requires money to be input or withdrawn only at the two instants 0 and T.

Therefore, an arbitrage possibility is a self-financing strategy h such that

A the value V of the strategy h at time 0 is zero: $V(h,0) = 0$;

B at future time T the strategy has positive value with probability 1: $V(h,T) > 0$ a.s.; that is the strategy is a machine that makes money.

We can summarise the main ideas on derivatives as follows:

- A financial derivative is defined in terms of some underlying that already exists on the market;

- The derivative must be priced in a way that is consistent with the underlying price given by the market in order to avoid mispricing between the derivative and the underlying;

- The price of the derivative is not obtained in some "absolute" sense, but it is determined in terms of the market price of the underlying asset.

Roughly speaking, let us assume to be at time t and to price a financial derivative with maturity T, $t < T$:

- In order to evaluate the derivative at t we have to estimate its future value;

- The value at T is a random variable X;

- The best prediction of X is the expected value conditional to the information available at t;

A derivative is evaluated according to the two following theorems of asset pricing.

Theorem 1 *First Fundamental Theorem of Asset Pricing. Consider the time interval $[0,T]$ and assume there exists a risk-free asset, and denote the corresponding risk-free interest rate by r.*

Then the market is arbitrage free if and only if there exists a probability distribution (martingale measure) Q such that

$$S(0) = \frac{1}{(1+r)^T} E_0^Q[S(T)],$$

where $S(t)$ is the asset price, $E_0^Q[\cdot]$ is the expected value under the measure Q, conditional to the information available at time 0. Then discounted prices are martingales under Q.

Proposition. In order to avoid arbitrage, a derivative, defined in terms of some underlying financial asset, must be priced according to the formula

$$V(S(0),0) = \frac{1}{(1+r)^T} E_0^Q[V(S(T),T)],$$

where $V(S(t),t)$ is the derivative value, $S(t)$ is the underlying price and Q is the martingale measure for the underlying market.

Theorem 2 *Second Fundamental Theorem of Asset Pricing. Assume that the model is arbitrage free. Then the market is complete if and only if the martingale measure is unique.*

There are several models for pricing derivatives, and a financial operator chooses the approach most suitable for his needs. The criteria used for the selection are:

- The ability to capture market characteristics and participants' behaviour;

- The inputs needed for model implementation, the ease of implementation itself and the accuracy of the implementation results;

- The cost and ease of maintaining the model operating.

In general, the steps for the derivative valuation process are:

- Observe the underlying market price behaviour and apply statistical tests to define its dynamics; therefore, assume the underlying dynamics.

- Define the derivative, create the model benchmark and test alternative models.

- Select the most appropriate model.

4.2 Closed Form Solution Models

A closed form solution model usually consists of finding the solution of a stochastic differential equation. The equation expresses the change in derivative value relative to all the key variables and it is subject to hedging assumptions and end conditions. The advantage of this kind of model is that they are very easy to implement; that is, given the inputs, the output is immediately obtained. The two main disadvantages are that the final formula is difficult to obtain and especially they are based on simplifying assumptions of the reality so that they do not reflect the real markets. The most known closed form solution model is the Black and Scholes [8] model for option pricing.

4.2.1 The Black and Scholes Model

We consider the time horizon $[t, T]$ and an economy with two assets:

- the risk free bond, $B(t)$, whose dynamics are described by the deterministic equation $dB(t) = rB(t)dt$;

- the stock, $S(t)$, which follows a stochastic behaviour according to a geometric Brownian motion $dS(t) = \alpha S(t)dt + \sigma S(t)dW(t)$.

We aim at pricing a third security which is a call option whose payoff at time T is

$$V(S(T), T) = max[S(T) - K; 0].$$

We look for the fair price $V(S(t), t)$ at time t. Black and Scholes [8] make the following reasonable assumption: as at T the price is just a function of the underlying price at T, $S(T)$, also at previous times t the price is a function of the underlying price, $S(t)$.

So we have the main following hypotheses:

- the price of the call is

$$c(t) = V(S(t), t); \tag{4.1}$$

- an investor can buy and (short) sell any quantities of all assets in the market, with no transaction costs;

- the market is free of arbitrage possibilities.

The stochastic dynamics of (4.1) is obtained by applying Ito's lemma[1] so that

$$dV(t) = \alpha_V(t)V(t)dt + \sigma_V(t)V(t)dW(t),$$

where

$$\alpha_V(t) = \frac{\frac{\partial V(t)}{\partial t} + \frac{\partial V(t)}{\partial S}\alpha S(t) + \frac{1}{2}\frac{\partial^2 V(t)}{\partial S^2}\sigma^2 S^2(t)}{V(t)}, \tag{4.2}$$

$$\sigma_V(t) = \frac{\frac{\partial V(t)}{\partial S}\sigma S(t)}{V(t)}. \tag{4.3}$$

In order to carry out a hedging strategy we can make a relative portfolio $h = (w_S, w_V)$ with relative weights w_S in the underlying and w_V in the derivative to replicate the bank account. The stochastic dynamics Π of the portfolio is

$$\frac{d\Pi(t)}{\Pi(t)} = \left\{ w_S \frac{dS(t)}{S(t)} + w_V \frac{dV(t)}{V(t)} \right\}, \tag{4.4}$$

that is

$$d\Pi(t) = \Pi(t)[w_S(t)\alpha(t) + w_V(t)\alpha_V(t)]dt + \Pi(t)[w_S(t)\sigma(t) + w_V(t)\sigma_V(t)]dW(t),$$

with the first constraint

$$w_S + w_V = 1. \tag{4.5}$$

The weights are chosen so that the diffusion part of this process vanishes by setting

$$[w_S\sigma(t) + w_V\sigma_V(t)] = 0. \tag{4.6}$$

Therefore, equation (4.4) becomes

$$d\Pi(t) = \Pi(t)[w_S(t)\alpha(t) + w_V(t)\alpha_V(t)]dt,$$

which implies, since the market is arbitrage free,

$$w_S(t)\alpha(t) + w_V(t)w_V(t) = r. \tag{4.7}$$

We aim at finding $V(S(t);t)$ such that condition (4.7) holds with constraints (4.5) and (4.6). We obtain

$$w_S = \frac{\frac{\partial V(t)}{\partial S}S(t)}{\frac{\partial V(t)}{\partial S}S(t) - V(t)}, \qquad w_V = \frac{-V(t)}{\frac{\partial V(t)}{\partial S}S(t) - V(t)}. \tag{4.8}$$

Now we substitute (4.2), (4.3) and (4.8) into condition (4.7), obtaining the Black and Scholes equation:

$$\frac{\partial V(t)}{\partial t} + \frac{\partial V(t)}{\partial S}rS(t) + \frac{1}{2}\frac{\partial^2 V(t)}{\partial S^2}\sigma^2 S^2(t) - rV(t) = 0. \tag{4.9}$$

---

[1] See Itô [45].

Therefore, finding the price of the derivative means finding $V(S(t),t)$ satisfying (4.9) with boundary condition

$$V(S(T),T) = f(S(T)) = max[S(T) - K, 0].$$

This is a parabolic partial differential equation that can be solved using the Feynman-Kač representation formula, that relates partial differential equations and stochastic differential equations. Under some technical conditions (which are satisfied in the Black and Scholes model) the solution $F(t,x)$ of the equation

$$\frac{\partial F(t,x)}{\partial t} + \frac{\partial F(t,x)}{\partial x} g(t,x) + \frac{1}{2} \frac{\partial^2 F(t,x)}{\partial x^2} h^2(t,x) S^2(t) + zF(t,x) = 0,$$

with boundary condition $F(X(T),T) = f(X(T))$ can be found as

$$F(t,x) = exp(z(T-t)) E_t[f(X(T))],$$

where X satisfies the stochastic differential equation

$$dX_u = g(u,x)du + h(u,x)dW_u.$$
$$X(t) = x$$

In the Black and Scholes model we have

$$X(T) = S(T),$$
$$g(t,x) = rS(t),$$
$$h(t,x) = \sigma S(t),$$
$$z = -r.$$

The solution $F(t,S(t)) = V(S(t),t)$ to the pricing problem is given by

$$V(S(t),t) = e^{-r(T-t)} E^Q(t)[f(S(T))], \qquad (4.10)$$

with $dS_u = rS_u du + \sigma S_u dW_u^Q$, $S(t) = s$. This is a risk-neutral valuation. Concluding:

- Technically, the fair price of a derivative is the discounted expectation of the payoff at maturity under a probability measure Q, that is the risk-neutral probability measure or martingale measure. The discounted prices are martingale under Q. Under Q the drift of the stock price process is equal to the risk free interest rate.

- The derivative price is consistent with the price of the underlying asset, therefore it is completely in line with the First Fundamental Theorem.

- The agents are not assumed to be risk neutral, but in any case the value of the derivative does not depend on the agents' preferences because it is calculated in a risk-neutral word.

Equation (4.10) becomes

$$V(s,t) = e^{-r(T-t)} \int_{-\infty}^{+\infty} max[se^Z - K, 0] f(z) dz,$$

where Z is a stochastic variable with the following normal distribution

$$N\left[\left(r - \frac{1}{2}\sigma^2\right)(T-t), \sigma\sqrt{T-t}\right],$$

and $f(z)$ is the corresponding density function.

After some standard calculations the Black and Scholes formula to price a European call option with strike price K and time of maturity T is

$$V(S(t),t) = sN[d_1(t,s)] - e^{-r(T-t)}KN[d_2(t,s)].$$

where $N[\cdot]$ is the cumulative distribution function for the $N[0,1]$ distribution and

$$
\begin{aligned}
d_1(t,s) &= \frac{1}{\sigma\sqrt{T-t}}\left\{log\left(\frac{s}{K}\right) + \left(r - \frac{1}{2}\sigma^2\right)(T-t)\right\}, \\
d_2(t,s) &= d_1(t,s) - \sigma\sqrt{T-t}.
\end{aligned}
$$

The price of the European put option with strike price K and time of maturity T is given by the formula $p(t) = V(S(t),t)$, where

$$V(S(t),t) = Ke^{-r(T-t)}N[-d_2(t,s)] - sN[-d_1(t,s)],$$

where, as before, $N[\cdot]$ is the cumulative distribution function for the $N[0,1]$ distribution and

$$
\begin{aligned}
d_1(t,s) &= \frac{1}{\sigma\sqrt{T-t}}\left\{log\left(\frac{s}{K}\right) + \left(r - \frac{1}{2}\sigma^2\right)(T-t)\right\}, \\
d_2(t,s) &= d_1(t,s) - \sigma\sqrt{T-t}.
\end{aligned}
$$

4.2.2 The Black Model

The Black [9] model is used to evaluate an option settled on a forward price. It derives from the Black and Scholes [8] model and the reader could follow the step of the pricing procedure of the previous section to obtain the following formula. The call option is written on a forward whose price is linked to the spot price according to the standard formula $F(t,T) = S(t)e^{r(T-t)}$, where $S(t)$ is the spot price, r the risk-free rate and T the maturity of the forward and the option. The forward price is lognormally distributed as the spot price. It has a deterministic dynamic $dF(t,T) = (\mu - r)F(t,T)dt$, where μ is the expected forward change rate. Therefore, the call price is

$$c(F(t,T),t) = e^{-r(T-t)}[F(t,T)N[d_1(t,s)] - KN[d_2(t,s)],$$

where $N[\cdot]$ is the cumulative standard normal distribution function and K the strike price. Furthermore,

$$d_1(t,s) = \frac{1}{\sigma\sqrt{T-t}}\left\{ log\left(\frac{F(t,T)}{K}\right) + \left(\frac{1}{2}\sigma^2\right)(T-t)\right\},$$

$$d_2(t,s) = d_1(t,s) - \sigma\sqrt{T-t},$$

where σ is the volatility of the spot price.

4.2.3 The Volatility Smile

Under the hypotheses of the Black and Scholes [8] model, the option underlying has a price volatility constant over time. This means that if we invert the formula of the option price in order to have the volatility for which the Black-Scholes price equals the market price, we obtain the so-called implied volatility of an option. Therefore, there is a one-to-one correspondence between prices and implied volatilities. The volatility implied in the Black and Scholes model can be obtained in the Excel environment by applying the solver tool. Furthermore, the option pricing model is built to incorporate the exact lognormal price distribution, so the volatility structure would be flat; that is, the same volatility would reflect all the options of the same expiration time but of varying strike prices. However, this does not actually occur! In fact if we determine the implied volatilities for options of the same time to expiration and on the same settlement price we obtain different values across different strike prices. Plotting the implied volatilities of options with the same characteristics and according to different strike prices we obtain a smile, like the one in Figure 4.1.

4.3 The Binomial Model

The binomial model was developed by Cox et al. [23]. They consider two points in time, $t = 0$ and a future data $t = 1$. There are three assets in the market:

- a risk-free asset: bond, with price $B(t)$ at time t,

- a risky asset: here we consider a stock, with price $S(t)$ at time t,

- a derivative: European call option, with price $c(t) = V(S(t),t)$ at time t.

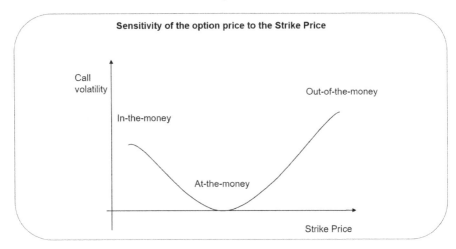

FIGURE 4.1: Volatility smile

The bond price evolves across time according to a deterministic process given by

$$B(0) = 1$$
$$B(1) = 1 + r, \text{ where } r \text{ is the risk-free rate.}$$

On the contrary, the stock price process is a stochastic process, whose behaviour is described as follows:

$$S(0) = s$$
$$S(1) = s\Omega, \tag{4.11}$$

where Ω is a random variable such that we have $\begin{cases} u, & \text{with probability } p_u \\ d, & \text{with probability } p_d \end{cases}$, where obviously $p_u + p_d = 1$. The term u is the incremental factor of the price, on the contrary d measures the price depreciation. Probabilities p_u and p_d are objective probabilities which depend on the expectation of the economic subjects.

At time 0 the following positive constants are known: $s, u, d, p_u, p_d, d < u$. The call value process is a stochastic process that depends on the stock process and whose dynamics are described by the following equations:

$$c(0) = V(S(0), 0),$$
$$c(1) = \begin{cases} c_u = max[0, us - K], & \text{with probability } p_u; \\ c_d = max[0, ds - K], & \text{with probability } p_d. \end{cases}$$

The assumptions of the model are:

- Short positions and fractional holdings are allowed.

- There is no bid-ask spread, i.e., the selling price is equal to the buying price of all assets.

- There are no transaction costs of trading.

- The market is completely liquid, i.e., it is always possible to buy or/and sell unlimited quantities on the market.

We define a portfolio $\omega = (x, y)$, where x is the holding quantity of bonds and y the stock quantity.

If x or y is negative this means that we are assuming a short position on the asset (we sell the asset); otherwise we assume a long position (we buy the asset).

The value process of the portfolio ω is defined by

$$\Pi(\omega, t) = xB(t) + yS(t), \quad t = 0, 1.$$

The model is free from arbitrage if and only if the following conditions hold:

$$d \leq (1 + r) \leq u. \tag{4.12}$$

In the binomial model it is very simple to obtain the risk neutral probability q_u and q_d. In fact, in an arbitrage free model the First Fundamental Theorem of Asset Pricing guarantees the existence of a martingale measure Q such that

$$s = \frac{1}{(1 + r)} E^Q[S(1)].$$

So we obtain

$$s = \tfrac{1}{1+r}[suq_u + sdq_d] \qquad q_u + q_d = 1, q_u, q_d > 0$$
$$q_u = \frac{(1+r)-d}{u-d}, \quad q_d = 1 - q_u = \frac{u-(1+r)}{u-d}.$$

Roughly speaking, the aim is to obtain a portfolio ω that replicates the values of the call option at time $t = 1$. That is possible because the market is complete. Consequently, because there are no arbitrage opportunities, the portfolio ω has also the same value of the option at time $t = 0$. In this way it is possible to obtain the pricing of the call option at time $t = 0$.

The portfolio ω which replicates call value at time 1:

$$\Pi(\omega, 1) = \begin{cases} c_u, & \text{if } \Omega = u; \\ c_d, & \text{if } \Omega = d, \end{cases}$$

is given by

$$x = \frac{1}{1+r} \cdot \frac{uc_d - dc_u}{u-d}, \quad y = \frac{1}{s} \cdot \frac{c_u - c_d}{u-d}. \tag{4.13}$$

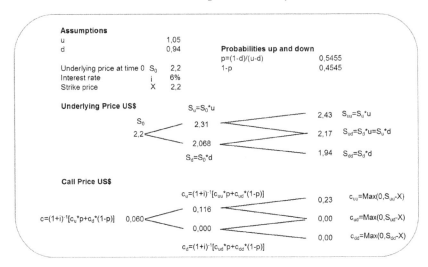

FIGURE 4.2: The tree scheme for the European option

At time 0 we have that $V(s,0) = \Pi(\omega,0)$, and using (4.13), we have

$$V(s,0) = x + sy = \frac{1}{1+r}\left[\underbrace{\frac{(1+r)-d}{u-d}}_{q_u} c_u + \underbrace{\frac{u-(1+r)}{u-d}}_{q_d} c_d \right].$$

So we have the risk neutral valuation formula

$$V(s,0) = \frac{1}{1+r}[c_u q_u + c_d q_d] = \frac{1}{1+r}E^Q[V(S(T),T)],$$

where Q is the martingale measure. Up to now we have shown the binomial procedure only when we assume two states of evolution for the stock price. We can sum up the binomial model through the well-known tree scheme. In Figure 4.2 we show the Excel implementation for the European call option pricing, whereas in Figure 4.3 the tree scheme for American options in Excel is illustrated. We recall that the American option can be exercised during the lifetime of the option and therefore, for example for a call, at each node of the tree, namely at time t, the payoff is $max[S(t) - K, c(t)]$, where $c(t)$ is the price of the European call option.

When we consider more than two points in time, we are defining a multiperiod binomial model. We can use a recursive formula for the stock price, so that at a generic time t it is:

$$S(t) = su^k d^{t-k}, k = 0, ..., t, \quad \text{with probability} \binom{t}{k} q_u^k q_d^{t-k}, \qquad (4.14)$$

where k denotes the number of up-moves that have occurred. This process can be represented by a binomial tree, such that each node can be represented by a pair

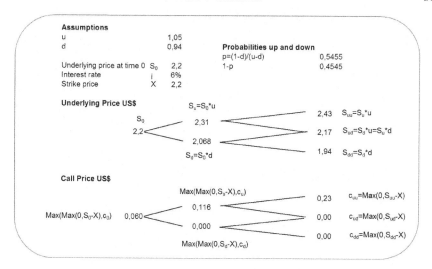

FIGURE 4.3: The tree scheme for the American option

(t,k). Then the option is priced through a backward procedure which means that we start from the maturity value of the option and proceed through nodes at previous times. If $\Pi(\omega,t,k)$ denotes the value of the portfolio at node (t,k) then it can be computed recursively by the scheme

$$\begin{cases} \Pi(\omega,t,k) = \frac{1}{1+r}[q_u\Pi(\omega,t+1,k+1)+q_d\Pi(\omega,t+1,k)], \\ \Pi(\omega,T,k) = max[0,su^kd^{T-k}-K]. \end{cases}$$

under the martingale probability previously seen and

$$x = \frac{1}{1+r}\frac{u\Pi(\omega,t,k)-d\Pi(\omega,t,k+1)}{u-d},$$

$$y = \frac{1}{S(t-1)}\frac{\Pi(\omega,t,k+1)-\Pi(\omega,t,k)}{u-d}.$$

In Figures 4.4 and 4.5 we show the pricing of a call option in the Excel environment. We divide the life time period of the option in more than two intervals and proceed according to the following steps:

1. Complete the tree of the stock price by applying formula (4.11) in the first approach in Figure 4.4, or formula (4.14) in the second approach in Figure 4.5;

2. Calculate the value of the option by starting from the value of option at maturity and proceeding backwards.

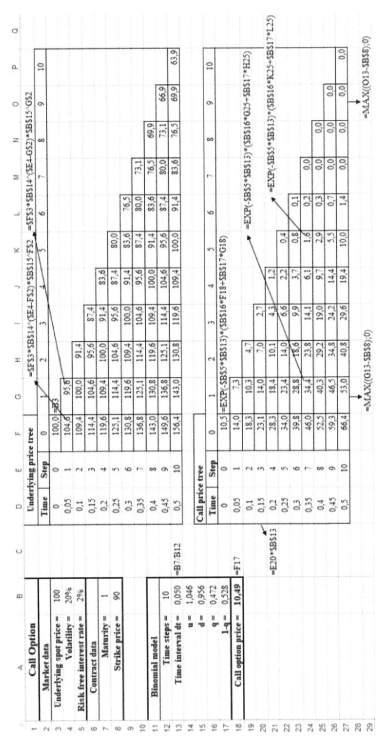

FIGURE 4.4: Excel implementation: Binomial model call option pricing, first approach

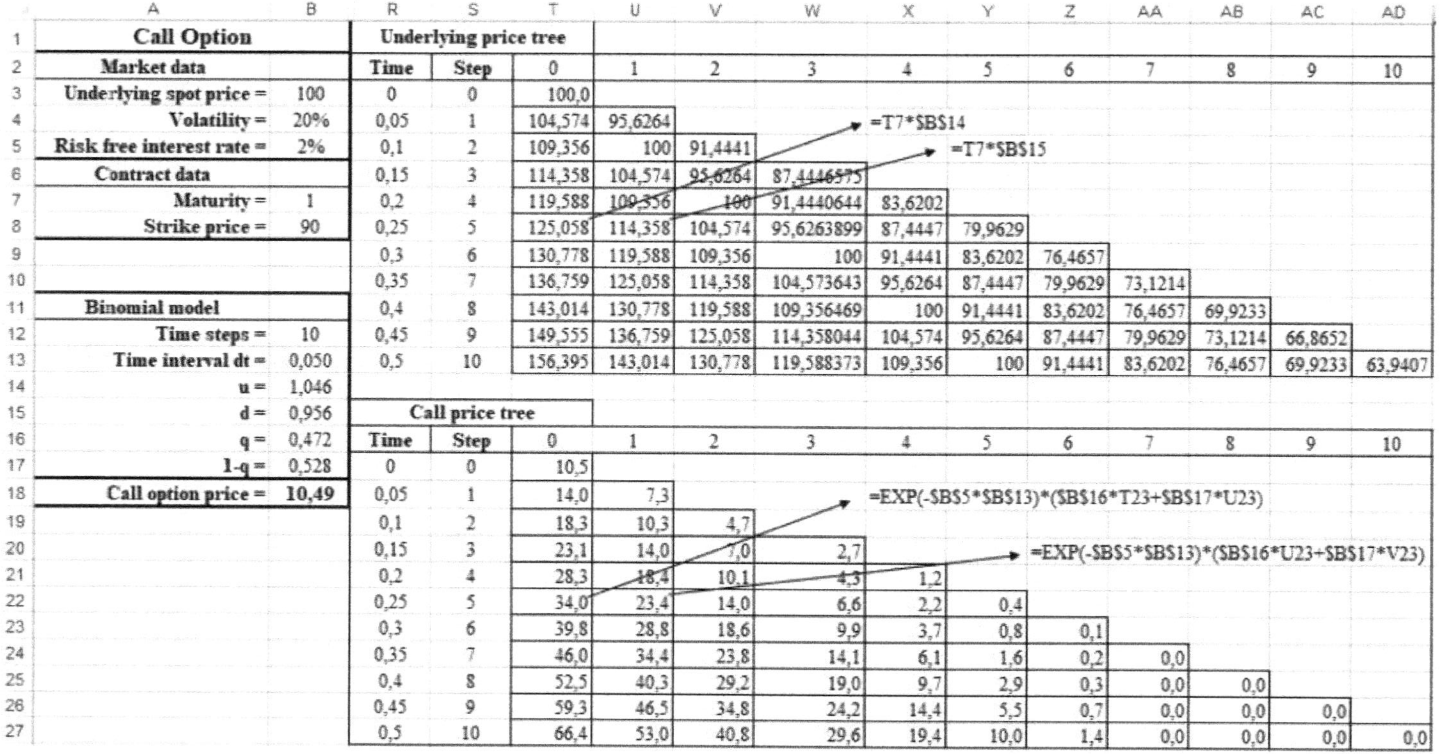

Call Option

Market data	
Underlying spot price =	100
Volatility =	20%
Risk free interest rate =	2%
Contract data	
Maturity =	1
Strike price =	90
Binomial model	
Time steps =	10
Time interval dt =	0,050
u =	1,046
d =	0,956
q =	0,472
1-q =	0,528
Call option price =	10,49

Underlying price tree

Time	Step	0	1	2	3	4	5	6	7	8	9	10
0	0	100,0										
0,05	1	104,574	95,6264									
0,1	2	109,356	100	91,4441								
0,15	3	114,358	104,574	95,6264	87,44655							
0,2	4	119,588	109,356	100	91,440644	83,6202						
0,25	5	125,058	114,358	104,574	95,626899	87,447	79,9629					
0,3	6	130,778	119,588	109,356	100	91,4441	83,6202	76,4657				
0,35	7	136,759	125,058	114,358	104,573643	95,6264	87,4447	79,9629	73,1214			
0,4	8	143,014	130,778	119,588	109,356988	100	91,4441	83,6202	76,4657	69,9233		
0,45	9	149,555	136,759	125,058	114,358044	104,574	95,6264	87,4447	79,9629	73,1214	66,8652	
0,5	10	156,395	143,014	130,778	119,588373	109,356	100	91,4441	83,6202	76,4657	69,9233	63,9407

=T7*$B$14   =T7*$B$15

Call price tree

Time	Step	0	1	2	3	4	5	6	7	8	9	10
0	0	10,5										
0,05	1	14,0	7,3									
0,1	2	18,3	10,3	4,7								
0,15	3	23,1	14,0	7,0	2,7							
0,2	4	28,3	18,4	10,1	4,3	1,2						
0,25	5	34,0	23,4	14,0	6,6	2,2	0,4					
0,3	6	39,8	28,8	18,6	9,9	3,7	0,8	0,1				
0,35	7	46,0	34,4	23,8	14,1	6,1	1,6	0,2	0,0			
0,4	8	52,5	40,3	29,2	19,0	9,7	2,9	0,3	0,0	0,0		
0,45	9	59,3	46,5	34,8	24,2	14,4	5,5	0,7	0,0	0,0	0,0	
0,5	10	66,4	53,0	40,8	29,6	19,4	10,0	1,4	0,0	0,0	0,0	0,0

=EXP(-$B$5*$B$13)*($B$16*T23+$B$17*U23)
=EXP(-$B$5*$B$13)*($B$16*U23+$B$17*V23)

FIGURE 4.5: Excel implementation: Binomial model call option pricing, second approach

4.4   The Monte Carlo Approach

The Monte Carlo method is widely used in finance to estimate the price of derivatives, or to represent the future behaviours of equities, bonds, exchange rates, portfolios, and so on. The basic idea is to simulate the dynamics of the asset across several scenarios in order to have at a future date different realizations of a random variable so that its expected value is a reliable estimation of the future asset market value. In practice, we repeat a random sample to calculate expectations, probabilities, variances and other estimates.

We now describe the Monte Carlo method for pricing a European call option, keeping in mind that the procedure is more or less the same as the pricing of other derivatives. Let us assume that the European call underlying spot price, $S(t)$, evolves according to a geometric Brownian motion. Through the Monte Carlo simulation the stochastic process representing the solution of the Brownian motion hypothesized for the underlying price is solved by a numerical tool. This makes it possible to estimate an approximate value for the expected value at maturity $E[S(T)]$ of the underlying. According to the First Theorem of Asset Pricing, the price of the option today, at time 0, will be equal to the discounted value at the risk-free rate r of the option value at maturity T and at a strike price K, $V(S(T),T) = max[0;S(T)-K]$. The steps of the procedure are:

1. The i-th time path of $S^i(t)$ is simulated under the risk-neutral probability measure;

2. The final value of the derivative is calculated, $V^i(S(T),T) = max[0;S^i(T)-K]$;

3. Steps 1 and 2 are repeated to obtain a fixed number Π of final sample values;

4. The expected final value of the derivative is estimated as the arithmetic mean of the final sample values:

$$V(S(T),T) = \frac{\sum_{i=1}^{\Pi} V^i(S(T),T)}{\Pi}; \qquad (4.15)$$

5. The current value of the derivative is calculated by discounting at the risk-free rate: $c(0) = V(S(0),0) = e^{-rT}V(S(T),T)$.

In Figure 4.6 we show the implementation of a call pricing through Excel software. The implementing procedure consists of the following steps:

1. Obtain standard normal draws for 500 simulations and 10 time steps;

2. Simulate the geometric Brownian motion over the 10 time steps and for the 500 times;

3. Calculate the call payoff at maturity for each simulated path of the geometric Brownian motion and discount it;

4. Average the discounted payoffs to obtain the Monte Carlo price.

Call Option

$=\$B\$3*EXP((\$B\$5-1/2*(\$B\$4^\wedge2))*\$B\$13+\$B\$4*F504*RADQ(\$B\$13))$

Market data		
Underlying spot price =	200	
Volatility =	20%	
Risk free interest rate =	5%	
Contract data		
Maturity =	0,5	
Strike price =	190	
Monte Carlo Approach		
Time steps =	10	
time interval dt =	0,05	=B7/B12
Simulation path number=	500	
Simulated price c₀ =	19,7677	= AVERAGE(P3:P503)

$=EXP(-\$B\$5*\$B\$7)*MAX((O3-\$B\$8);0)$

	Simulated path S^i_t	Time steps t				Simulated call price at maturity $V(S^i_T,T)$
		1	2	3	10	
1		200,1989	197,3136	201,7829	180,8916	0,0000
2		203,8133	209,2635	215,3365	202,2057	11,9043
3		180,6504	163,7602	167,5090	188,3611	0,0000
4		193,9520	197,5803	189,0790	180,7126	0,0000
5		192,6567	190,3107	182,2558	206,7797	16,3654
6		194,1451	186,6427	169,4512	172,0069	0,0000
7		194,7531	202,6104	204,0600	167,3252	0,0000
8		190,0516	188,5901	188,5341	217,8410	27,1536
9		183,2791	177,6850	191,3052	201,1995	10,9230
10		199,4120	184,3911	180,9220	171,7866	0,0000
11		204,3538	199,6214	193,9958	213,4798	22,9000
12		181,7202	180,0835	182,9926	193,0581	2,9826
13		195,3779	194,9787	198,3978	175,6863	0,0000
14		206,5277	245,6810	255,4231	279,3327	87,1271
15		205,7089	205,4701	201,1500	204,8199	14,4540

	Normal standard sample	1	2	3	10	= NORM.S.INV(RADN())
1		-0,011318	-0,35815	0,4672979	-0,888392	
2		0,3887843	0,5565596	0,6061434	-2,456961	
3		-2,308823	-2,228475	0,4725715	-0,457632	
4		-0,720159	0,3808929	-1,016969	-1,528675	
5		-0,869993	-0,30751	-1,000562	-0,483283	
6		-0,697909	-0,914768	-2,194281	-0,569464	
7		-0,62799	0,85508761	0,1258645	0,1553993	

FIGURE 4.6: Excel implementation: Call option pricing

Figure 4.7 contains the main formulas to price an Asian option in Excel, according to the following steps:

1. Obtain standard normal draws for 500 simulations and 10 time steps;

2. Simulate the geometric Brownian motion over the 10 time steps and for the 500 times;

3. Calculate the discounted Asian call payoff at maturity for each simulated path of the geometric Brownian motion according to the formula:

$$V(S^i(T), T) = e^{-rT} \frac{\sum_{t=1}^{10} S^i(t)}{10}; \qquad (4.16)$$

4. Average the discounted payoffs to obtain the Monte Carlo price.

4.5   Exercises

1. Assume that the actual underlying price is $25, its volatility is 30% and drift equal to 4%. Furthermore, the annual risk-free interest rate in the market is 3%. Price a European call option with maturity 6 months and strike price $25 according to the Black and Scholes model.

2. Apply the put-call parity relationship to obtain the European put price corresponding to the call price of Exercise 1.

3. Consider an underlying commodity with spot price $45 and expected return of 6%. Use the Excel Solver to find the implied volatility of an European call option with maturity one year and strike price $50.

4. Consider two trading dates $t = 0, 1$ and a commodity with price at time 0 equal to $50. Given the interest rate is 3%, the strike price is $50 and $u = 1.1$ and $d = 0.97$, determine the European call and put option prices through the binomial model.

5. Consider three trading dates $t = 0, 1, 2$ and a commodity with price at time 0 equal to $45. Given the interest rate is 3%, the strike price is $50 and $u = 1.1$ and $d = 0.97$, determine the European call option price through the binomial model.

6. Consider three trading dates $t = 0, 1, 2$ and a commodity with price at time 0 equal to $50. Given the interest rate is 3%, the strike price is $50 and $u = 1.1$ and $d = 0.97$, determine the American call and put option prices through the binomial model.

	A	B	C	E	F	G	H	O	P	Q	R	S
1	**Asian Option**				**Time steps t**	=$B$3*EXP(($B$5-1/2*($B$4^2))*$B$12+$B$4*F504*RADQ($B$12))						
2	**Market data**			**Simulated path S^i_t**	1	2	3			10	**Simulated price at maturity $V(S^i_T,T)$**	
3	**Underlying spot price =**	100		1	100,1827	99,7198	98,3599	98,3031	111,4647	98,2898	0,0000	= EXP(-$B$6*$B$7)*MAX((Q3-AVERAGE(F3:P3));0)
4	**Volatility =**	20%		2	103,0257	97,3066	97,9828	86,8409	81,3458	96,5648	4,8341	
5	**Risk free interest rate =**	2%		3	88,4921	86,9108	83,2635	61,8707	63,8489	71,4974	0,0000	
6	**Contract data**			4	100,9219	105,4846	98,7540	87,4303	85,3162	77,3286	0,0000	
7	**Maturity =**	1		5	96,5449	96,4092	96,3133	85,3065	89,2886	85,3906	0,0000	
8				6	95,8862	90,7850	98,8392	71,6270	74,3422	75,1000	0,0000	
9				7	102,5218	102,5339	103,4909	80,6369	75,3668	77,4393	0,0000	
10	**Monte Carlo Approach**			8	101,8138	96,4750	100,4062	113,2320	105,9484	105,1402	0,2591	
11	**Time steps =**	10		9	99,6491	106,3594	104,1693	97,6446	95,5877	100,1297	0,0000	
12	**time interval dt =**	0,083	=1/12	10	91,5144	95,8080	98,8056	67,7269	63,0020	70,6835	0,0000	
13	**Simulation path number=**	500		11	97,9815	94,9462	81,8119	86,5555	93,2705	79,7775	0,0000	
14	**Simulated price c_0=**	3,6742	= AVERAGE(P3:P503)	12	107,6710	106,9579	105,9423	141,9884	164,1111	123,1669	0,0000	
15				13	98,4206	97,9562	100,1123	81,1197	83,3998	74,4183	0,0000	
16				14	105,0135	98,7925	108,4742	177,2939	176,3397	168,0159	32,2740	
17				15	96,9683	94,5590	100,9698	111,6708	112,2244	104,4381	1,2546	
502												
503				Normal standard sample	1	2	3			10		
504				1	0,031615	-0,08022	-0,23783	-0,62044	2,1763658	-0,6228	= NORM.S.INV(RADN())	
505				2	0,516288	-0,9892	0,119961	-0,5674	-1,132216	1,270943		
506				3	-2,11755	-0,31232	-0,74255	-0,33905	0,5450981	2,165722		
507				4	0,158951	0,76588	-1,14199	0,025476	-0,423971	-2,1011		
508				5	-0,60902	-0,02436	-0,01724	-0,85194	0,7902181	-0,83487		
509				6	-0,72761	-0,94687	1,472251	-1,23248	0,6444335	-0,41239		
510				7	0,431376	0,002034	0,160919	0,779017	-1,170684	0,078194		

FIGURE 4.7: Excel implementation: Asian option pricing

7. Consider a European call option with maturity 6 months and an underlying commodity with price at time 0 equal to $45 and volatility 20%. Given the interest rate is 3%, the strike price is $50 and $u = 1.1$ and $d = 0.97$, determine the European call option price through the binomial model by implementing 10 tree time steps.

8. Consider a European call option with maturity 6 months and an underlying commodity with price at time 0 equal to $100, volatility 30% and expected return 5%. Given the interest rate is 3%, the strike price is $90. Determine the European call option price through the Monte Carlo simulation by implementing 10 tree time steps and 500 simulations.

4.6   Answers

1. $2.29.

2. $1.92.

3. 25.11%.

4. $2.24; $0.78.

5. $0.89.

6. $3.67; $0.80.

7. $11.99.

8. $\approx$ $10.

Chapter 5

Applications

5.1 Modelling the Italian Electricity Spot Market

The Italian spot electricity price, called PUN (Prezzo Unico Nazionale), is a market equilibrium price, and it is formed by the intersection of the supply and demand curves. This pricing process ensures that prices are characterised by an uncertainty bigger than the one existing in other financial markets. Furthermore, electricity underlies transportation constraints, its demand is very inelastic, and the market is strongly affected by the Italian power plant park. Thus, before addressing a theoretical model for the daily electricity spot price, the price time series may give us a picture of the basic price pattern and a first idea about appropriate modelling approaches[1]. In Figure 5.1 we plot the daily time series of the baseload electricity price in Italy, that is the PUN, from January 1, 2012 to July 6, 2015.

---

[1] We thank the engineer Luana Bruno for her help with the data analysis of this section.

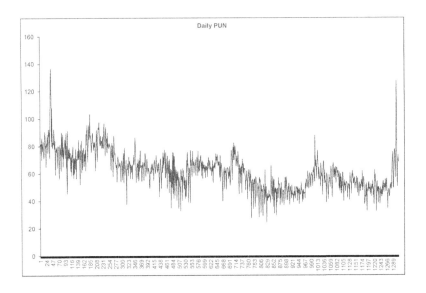

FIGURE 5.1: Italian power spot price: daily time series 01.01.2012 - 07.06.2015

Electricity prices show the following characteristics:

1. presence of jumps,

2. high volatility,

3. mean reversion,

4. seasonality,

5. asymmetric probability distribution.

Jumps or Spikes

The electricity spot price is often subject to sudden movements upwards or downwards, forming jumps with respect to a mean value. These jumps are also called spikes. However, the price tends, in a short time, to return to the previous level. The higher the jump, the greater the probability that the price will return to its average level. Jumps tend to concentrate in certain hours of the day and the week, which is when energy demand increases and production capacity is at its maximum. The spikes are related to the particular characteristic of nonstorability. Indeed, unlike the other primary energy sources, electricity is impossible to store, with the exception of hydropower, for technical reasons. Therefore, the electricity must be consumed as soon as it is traded. Hence, excessive demand or supply causes the price to be very volatile.

High Volatility

The volatility measures the percentage price change over time, and it is calculated as the standard deviation of returns. Power volatility appears to be much higher than that of other commodities or other financial assets. Crude oil and natural gas have a volatility ranging between 1.5% and 4%, the zero-coupon bond volatility is less than 0.5% and stocks and corporate bonds are very volatile, with a volatility greater than 4%. For electricity, however, the volatility can reach the level of 30%. The reasons for high volatility are several: the nonstorability, the effect of a highly inelastic demand, and the fact that electricity consumption is both tied to weather conditions and constrained by technical limitations of transportation and production. Another aspect that is evident in price patterns (see for example Figure 5.1) is the persistence of volatility (i.e., volatility clusters); indeed price remains at a mean level for long periods outlining precisely the areas with high volatility and the areas with low volatility.

Mean Reversion

The mean reversion is the tendency of prices to return to their long-term average value. Therefore, sudden spikes are immediately followed by jumps in the opposite direction. The speed of mean reversion depends on a number of factors including the presence of periodicity in prices and the kind of technologies adopted for the production and distribution of energy.

Seasonality

Electricity spot price time series often show monthly, weekly and intra-day seasonality. The monthly seasonality is due to the effects of weather conditions on energy demand, which can vary from country to country. There are countries where energy demand is greatest in the winter because of low temperatures and other countries where the peak demand is in the summer months due to the high temperatures. The weekly periodicity is caused by the difference between the power demand on weekdays on one side and the demand on weekends or on public holidays (when companies do not work) on the other side. This last demand appears to be significantly lower than the weekday demand, thus, consequently, prices are low. In Figure 5.2 we show the average hourly prices on different days of the week. We can observe that the price curve on Saturday is higher than the price curve on Sunday, and the curve referring to the hours between midnight and 6 a.m. is even higher than the price curve on weekdays. The intra-day seasonality, instead, depends substantially on people's daily life style: for example, on weekdays the price begins to increase at 6 a.m., when people wake up, and reaches the first peak at 9 a.m. (opening time for offices and commercial activities) and there is a second peak after 6 p.m.

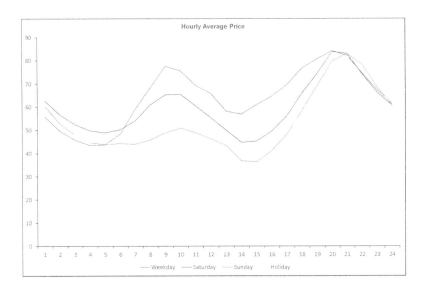

FIGURE 5.2: Hourly average price per day

Asymmetric Probability Distribution

The electricity price has a probability distribution that is positively asymmetric and leptokurtic. There is a positive skewness because the distribution has a long tail to the right, and this is due to the fact that most jumps are upwards, probably because of sudden shocks in demand. The leptokurtic distribution has fat tails; that is, there is an excess of kurtosis, also due to the presence of jumps. Kurtosis is much more marked in electricity prices than in other commodities.

5.1.1   Data Analysis

We analyse the time series of the Italian daily spot price of electricity, PUN, traded on the Italian Exchange, from January 1, 2012 to July 31, 2015, for a total of 1,308 observations. The PUN is determined as the average daily price per hour of every day. After a preliminary qualitative analysis by observing the time plot in Figure 5.1, we individuate the following price characteristics:

• High frequency of observations: prices are always provided at constant hourly time intervals (a price for each hour of the day for every day of the year);

• Presence of jumps: we can observe a peak of 132.48 €/MWh on February 9, 2012, due to the wave of frost that gripped much of Europe for more than

TABLE 5.1: Descriptive statistical analysis of PUN daily time series, 01.01.2012 - 07.06.2015

Mean	61.731
Standard Error	0.378
Median	61.614
Standard Deviation	13.7 %
Kurtosis	1.487
Asymmetry	0.638
Minimum	24.613
Maximum	136.665
Jarque-Bera Test	213.578

fifteen days and a peak of 127.46 €/MWh on July 23, 2015, when in Italy the weather temperature went up to $41^{o}C$;

• Mean reversion: due to sudden jumps that go back fairly quickly to the previous level;

• Seasonality: it is difficult to identify the magnitude of seasonality by observing Figure 5.1, but from Figure 5.2 we can deduce that electricity price exhibits weekly periodicity.

5.1.2 Price Analysis

In order to verify price features that emerged from the qualitative analysis of the time series of Figure 5.1, we need to make a descriptive analysis.

In Table 5.1 we summarise the results of PUN time series statistical analysis. The probability distribution has a slight positive skewness; indeed the right tail is slightly more elongated then the left one, and has an excess kurtosis, that is it shows fat tails. Furthermore, the PUN is characterised by a high volatility, of about 13.7 %, and the mean production cost of the Italian electric park is about 65 €/MWh.

We now want to investigate the deviation of PUN probability distribution from a normal distribution. The value of the Jarque-Bera statistic in Table 5.1 states that the hypothesis of normality for PUN density is rejected. Indeed, the analysis of price data reveals a probability for an "extreme" event higher than the one predicted by the normal distribution. However, for electricity spot prices, the deviation from normality is more evident than equities and most other types of commodities. In Figure 5.3 we show the histogram of PUN data, whereas in Figure 5.4 we show the QQ plot as normality tests. If empirical returns are normally distributed, we expect to observe the red curve corresponds with the dashed straight line in Figure 5.4. However, this is not the case. On the contrary, we find clear indication for fat tails and that is that "extreme" events occur more often with respect to what is predicted by a normal distribution.

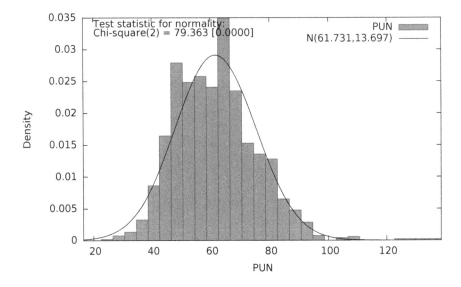

FIGURE 5.3: PUN price histogram

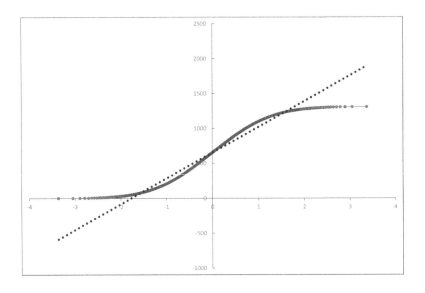

FIGURE 5.4: QQ plot for PUN

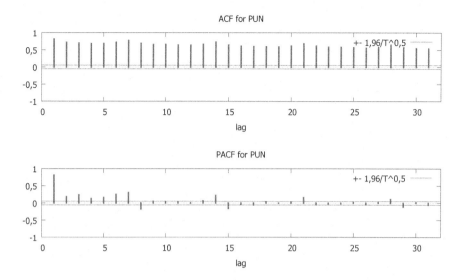

FIGURE 5.5: Autocorrelation function and partial autocorrelation function for PUN

Furthermore, if we plot the autocorrelation function and the partial autocorrelation function[2] in Figure 5.5, we observe that there is a strong correlation among data, in particular the time series has an autoregressive character due to a seasonality of seven periods (that is, a weekly seasonality).

In order to investigate the seasonality of the electricity spot price, we analyse the daily volatility of log-returns, calculated by using hourly prices of each day. Thus, in Figure 5.6 we show a volatility for each date of our time series (2012/01/01-2015/07/31), that is, for each day of the week.

Over the considered time period, the average volatility is 13.7%, whereas the peak volatility is reached on 2013.05.01, due to a sudden PUN reduction of 4% at 2.00 p.m. and a subsequent increase of about 4% at 3.00 p.m., compared with an average daily price change of $\pm$ 2%. At certain hours of the day, for example at 7.00 a.m., 2.00 p.m. or 9 p.m., the price is more volatile than at other hours, as we show in Figure 5.7. Furthermore, on certain days of the week, on average, the volatility is higher than on others, giving rise to volatility clusters. In Figure 5.8 we observe that the highest volatility occurs on Sunday.

The reasons why the volatility of electricity prices is so high are related to several factors, including the uncertainty of demand, the variation of the production due to weather conditions, the uncertainty in generating electricity (risk of outages) and, finally, the network congestion and market power.

---

[2]See Chapter 6 for an explanation of the statistical tools.

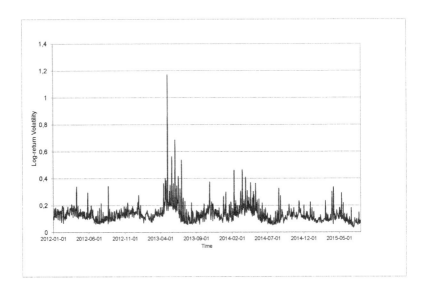

FIGURE 5.6: Daily log-return volatility

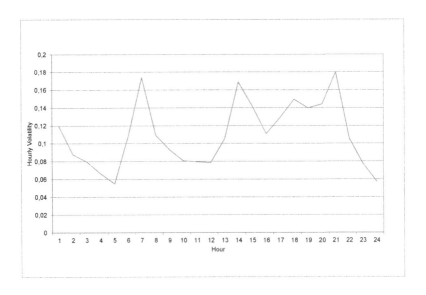

FIGURE 5.7: Hourly log-return volatility

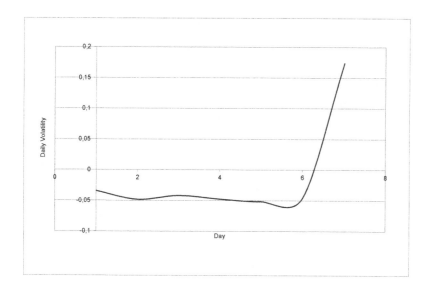

FIGURE 5.8: Daily log-return volatility over the week

5.1.3 Log-return Analysis

After having removed the yearly and weekly seasonality, we analyse the characteristics of the logarithmic return time series. In Table 5.2 we show the descriptive statistical analysis of the time series. We find that the probability distribution has a slight positive skewness and an excess kurtosis - leptokurtic distribution, because it has fat tails due to the presence of jumps.

In addition, the standard deviation, namely the volatility, is equal to 14.8%.

TABLE 5.2: Descriptive statistical analysis of PUN log-return daily time series, 01.01.2012 - 07.06.2015

Mean	0.00
Standard Error	0.004
Median	-0.003
Standard Deviation	14.8%
Kurtosis	3.288
Asymmetry	0.446
Minimum	-0.591840133
Maximum	0.707
Jarque-Bera Test	48.015

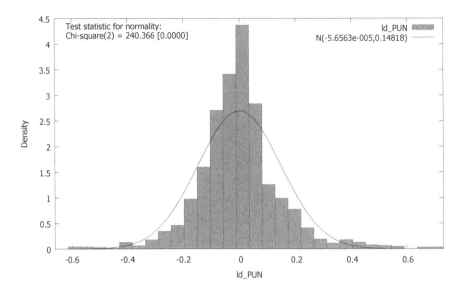

FIGURE 5.9: PUN log-return histogram

We can carry out normality tests by plotting the histogram of the log-returns and the QQ test output in Figures 5.9 and 5.10.

Consequently, we argue that electricity spot log-return does not distribute according to a Gaussian distribution because too many "extreme" values exist. Therefore, we remove these "extreme" values, that are the spikes, from the original time series by using a numerical procedure that recursively filters returns that are greater than three times the standard deviation. This procedure is described in detail in Section 5.1.4.

After having removed the spikes from the original time series, we find that the fat tails nearly disappeared and the deviation of the empirical returns from returns predicted by a Gaussian distribution becomes negligible, as revealed by normality tests in Figures 5.11 and 5.12.

Overall, if we consider our total descriptive analysis, we can argue that electricity prices exhibit seasonality and are characterised by spikes and mean reversion.

5.1.4   The Model

We assume a finite time horizon $\overline{T}$, where the uncertainty is represented by a filtered probability space $(\Omega, \mathscr{F}, (\mathscr{F}_t)_{t \geq 0}, \mathbb{P})$. $\mathscr{F} = \mathscr{F}_{\overline{T}}$ is the σ-algebra at time $\overline{T}$. All statements and definitions are understood to be valid until the time horizon $\overline{T}$.

We recall that the data analysis of previous sections reveals three distinctive characteristics of electricity prices which should be represented in a pricing model: mean reversion, spiky behaviour and seasonality. In order to capture all the described features, we use a mean reversion jump diffusion model, that can be considered a jump-augmented model of Schwartz [60]. If we indicate with $(S(t))_{t \geq 0}$ the electricity spot

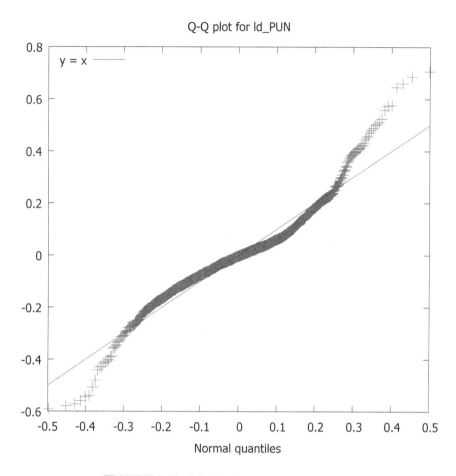

FIGURE 5.10: QQ Plot for PUN Log-returns

price process, we set the log-price as $x(t) = lnS(t)$, and the model under a risk-neutral measure $\tilde{\mathbb{P}}$ is defined as follows:

$$dx(t) = \alpha(\widehat{\mu} - x(t))dt + \sigma dW(t) + kdZ(t), \qquad (5.1)$$

where α indicates the speed of mean reversion of the price $S(t)$ towards a long-run mean $e^{\widehat{\mu}}$. The long-run parameter $\widehat{\mu}$ takes into account the market price of risk, in the hypothesis of a risk neutral world and $W(t)$ the Brownian motion under the risk neutral measure. Furthermore, $(J(t))_{t \geq 0}$ is the jump process, a discrete time process, i.e., Poisson process. The annualized frequency of jumps is given by the average of jumps per year, which we call ζ. The jump return size is k, which is determined by the natural logarithm of the jump returns being normally distributed, $\ln(1 + k) \sim N(\ln(1 + \bar{k}) - 1/2\gamma^2, \gamma^2)$, where $\bar{k}$ is the mean jump size and γ is the standard deviation.

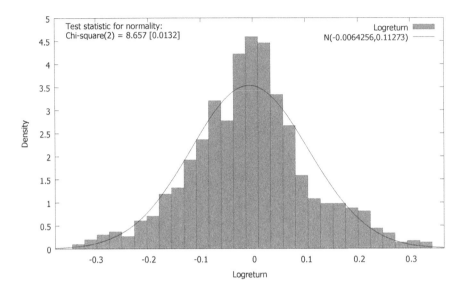

FIGURE 5.11: PUN log-return histogram without spikes

We estimate parameters of mean reversion in equation (5.1) by using the deseasonalized data set of log-return without jumps, obtained in Subsection 5.1.4. In order to verify the mean reversion, we apply the Augmented Dickey Fuller test (ADF) and the variance ratio test (see Cochrane [22] and Lo and MacKinlay [51], among others). The ADF gives an asymptotic p-value of 1.049e-005, meaning that the time series is stationary. The variance ratio test verifies if the log-price follows a random walk, a trending dynamic or a mean-reverting process. According to the test, we study the variability of returns over different horizons, relative to the variation over a one-day period. We find that the variance ratio statistic is always less than one, meaning that the data-generating process of time series is mean-reverting. We plot the variance ratio as a function of the time horizon in Figure 5.13.

In order to obtain estimates of mean reversion parameters in (5.1), we have to consider only the mean-reverting part of the process in discrete time:

$$\Delta x(t) = \alpha(\widehat{\mu} - x(t))\Delta t + \sigma\sqrt{\Delta t}\varepsilon(t), \qquad (5.2)$$

where $\Delta x(t) = x(t+1) - x(t)$, Δt is the unit time interval, and $\varepsilon(t) \sim N(0, \Delta t)$ is a normally distributed random variable. The discrete-time mean reversion model is equivalent to the following simple linear model:

$$\Delta x(t) = \alpha_0 + \alpha_1 x(t) + \sigma\varepsilon(t), \qquad (5.3)$$

where $\alpha_0 = \alpha\widehat{\mu}\Delta t$, $\alpha_1 = \alpha\Delta t$ and $\varepsilon(t) \sim N(0, \Delta t)$. Parameters in (5.2) are estimated by applying the OLS regression to (5.3) and the results are shown in Table 5.3.

The parameters of the jump process are estimated by implementing a jump-recursive filter estimation procedure already mentioned in Subsection 5.1.3. This

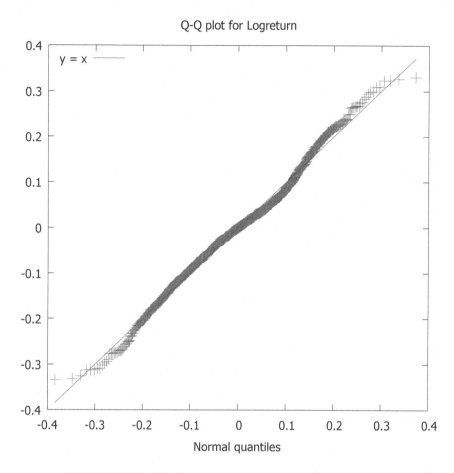

FIGURE 5.12: QQ plot for PUN log-returns without spikes

procedure attempts to identify and characterise the lower frequency, the higher mean jump size and jump return. It consists of the following steps:

1. Calculate sample return standard deviations;

2. Identify returns greater than a fixed threshold; we choose three or four standard data deviations, but it depends on market data; we choose a threshold of three standard deviations and we call these values "jumps";

3. Remove the jumps from the sample;

4. And repeat until convergence in jump intensity ξ, mean jump size $\bar{k}$ and jump volatility γ.

The results of the iterated procedure are summarised in Table 5.4.

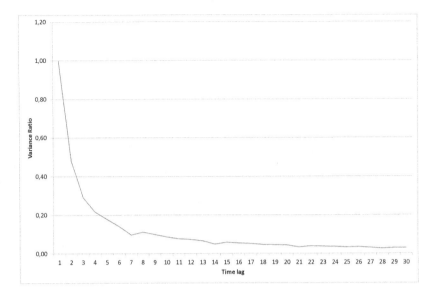

FIGURE 5.13: Variance ratio function for no-jump log-returns

The jumps have both positive sign and negative sign, therefore, the average amplitude of the jump cannot be defined as the jump average, because jumps of opposite sign cancel, but we have to calculate average jump amplitudes separately for the positive case and the negative one. We then obtain two different probability distributions and we calculate different parameters. The parameter estimates are given in Table 5.5:

Finally, the last feature of electricity prices that has to be captured is the seasonality. We incorporate seasonality in the volatility σ of the process (5.1). In Subsection 5.1.2 we have investigated the weekly seasonality of electricity prices. In particular we have observed a periodic variability of prices that varies according to the different

TABLE 5.3: Regression results for equation $\Delta x(t) = \alpha(0) + \alpha_1 x(t) + \sigma \varepsilon(t)$

	Coefficient	Std. Error	t-ratio	p-value
$\alpha(0)$	0.579970	0.0794474	7.3000	0.0000
α_1	0.858453	0.0155550	55.1884	0.0000

Residual Std Error	0.136612
R^2	0.621947
Adjusted R^2	0.621657
F-statistic	3666.524
P-value(F)	0.00

TABLE 5.4: Parameter estimates from jump-recursive filter estimation procedure

Iteration	Return Std Dev	Jump threshold	No. Jumps	ξ	$\bar{k}$	γ
0	0.1482	0.4445	-	-	-	-
1	0.1284	0.3851	25	6.9816	0.1949	0.5244
2	0.1181	0.3542	46	12.8462	0.1317	0.4779
3	0.1142	0.3427	55	15.3596	0.1446	0.4544
4	0.1131	0.3393	58	16.1974	0.1311	0.4524
5	0.1127	0.3382	59	16.4767	0.1347	0.4493
6	0.1127	0.3382	59	16.4767	0.1347	0.4493

TABLE 5.5: Parameter estimates for positive and negative jumps

	No. Jumps	ξ	$\bar{k}$	γ
Positive Jump	38	10.6120	0.4587	0.0975
Negative Jump	21	5.8645	-0.4518	0.0834

days of the week. Therefore we define a volatility that assumes a different value for each day of the week as follows:

$$\sigma = \sigma(0) + \sum_{i=1}^{6} \sigma_i D_i, \tag{5.4}$$

where $\sigma(0)$ is the volatility coefficient of Sunday that is reduced by the other coefficients σ_i, $i = 1, ..., 6$, in order to have the volatilities of the other weekdays. In particular σ_i represents the volatility parameter of the *i-th* day of the week, with $i = 1$ for Monday, $i = 2$ for Tuesday and so on until $i = 6$ for Saturday, whereas D_i, $i = 1, ...6$ are dummy variables[3]. Parameters of equation (5.4) are estimated by the OLS regression with dummies. The results are displayed in Table 5.6.

5.1.5 Hedging Strategy

We refer to an intermediary company that buys electricity from a producer and sells it to the final consumer. This kind of company is clearly exposed to the price risk due to the high variability of electricity prices. Furthermore, we assume that the market expects a decrease of the electricity price. In order to hedge this risk, the company stipulates a forward contract with the final consumer to lock in the power selling price at a specific future date, and it buys a call option to protect against an increase of the purchase price from the producer side.

In electricity markets it is used to replace the plain vanilla options with the so-called Asian options (see Eydeland and Geman [28], Lucia and Schwartz [52] and

---

[3]Fanelli et al. [32] use a similar seasonal volatility function to model dynamics of German futures prices.

TABLE 5.6: Regression results for equation $\sigma = \sigma(0) + \sum_{i=1}^{6} \sigma_i D_i$

	Estimate	Std Error	t value	p-value
Constant/Sunday	0.173872	0.004621	37.624	<2e-16
Monday	-0.03434	0.006535	-5.254	1.73e-07
Tuesday	-0.048289	0.006535	-7.389	2.64e-13
Wednesday	-0.041794	0.006535	-6.395	2.24e-10
Thursday	-0.047635	0.006535	-7.289	5.42e-13
Friday	-0.051578	0.006535	-7.892	6.27e-15
Saturday	-0.046194	0.006544	-7.059	2.73e-12

Residual Std Error	0.05583
R^2	0.2738
Adjusted R^2	0.2738
F-statistic	37.54
P-value(F)	< 2.2e-16

Weron [63], among others). An Asian option has a path-dependent payoff because it depends on the behaviour of the underlying price during all option lifetime or part of it. More precisely, the payoff at maturity depends on the average underlying price calculated over a pre-set period of time. In the following, we indicate this pre-set period as the "payoff period". The Asian options are particularly suitable for developing strategies in power markets because electricity contracts have the characteristic that the commodity delivery occurs over a time interval, instead at a fixed date, so that it is appropriate to consider a derivative instrument that captures price behaviour over a pre-set period.

The average price can be influenced by several factors, including:

• The time interval used to calculate the average of prices;

• The type of average, which may be arithmetic or geometric;

• The kind of sample we use; it may consist of discrete or continuous data.

We focus on arithmetic Asian options and we can distinguish between two types of them:

• The average price options, whose final underlying settlement price is calculated as an average price over the payoff period. In this case, the payoff $\phi(T)$ of an Asian call at maturity T and strike price K is:

$$\phi(T) = max[\frac{1}{n} \sum_{j=1}^{n} S(j) - K; 0], \qquad (5.5)$$

where $S(j)$, $j = 1, ..., n$ are the n prices observed over the payoff period.

TABLE 5.7: Hedging strategy cash flows for the electricity intermediary company. The strategy consists of a short position on a forward contract with maturity one year and a long position on an arithmetic Asian call option with the same maturity T. $V(0)$ indicates the Asian call premia. F is the forward price.

Time	Forward	Asian Option
0	0	$-V(0)$
T	$S(T) - F$	$\phi(T)$

- The average strike option, whose strike price is given by the average price over the payoff period. In this case, the payoff $\phi(T)$ of an Asian call at maturity T is:

$$\phi(T) = max[S(T) - \frac{1}{n}\sum_{j=1}^{n} S(j); 0], \qquad (5.6)$$

where the strike price is given by the average value $\frac{1}{n}\sum_{j=1}^{n} S(j)$ and $S(j)$, $j = 1,...,n$ are the n prices observed over the payoff period.

The considered hedging strategy consists of a short position on a forward contract with forward price F and maturity one year T, and a long position on an arithmetic Asian call option with the same maturity T and strike price K. We consider an average price option and an option period equal to one year in a first case and one month in a second case. The cash flows of the strategy are summarised in Table 5.7.

As we show in Table 5.7, at inception, the value of the forward contract is null, whereas the Asian call option value is $V(0)$, that represents the premium of the option. At maturity T the value of the forward contract is given by the difference between the price cashed in and the market spot price $S(T)$, whereas the payoff of the Asian call is given by formula (5.5).

The Asian call premium is obtained by discounting the expected payoff at maturity T, conditional to the information available at initial time 0, $\mathscr{F}(0)$, calculated under the risk-neutral probability measure Q, as follows:

$$V(0) = e^{-\delta T} E^{Q}[\phi(T)|\mathscr{F}_0], \qquad (5.7)$$

where $e^{-\delta T}$ is the discount factor at a risk-free rate δ.

In order to evaluate the option at inception, we assume three scenarios according to three different strike prices that are fixed in the contract:

(a) Scenario "In-the-money": the strike price is lower than the electricity spot price in the market at inception;

(b) Scenario "At-the-money": the strike price is equal the electricity spot price in the market at inception;

(c) Scenario "Out-of-the-money": the strike price is higher than the electricity spot price in the market at inception.

TABLE 5.8: Case 1: Results from the Monte Carlo estimation procedure. We estimate the premium of an Asian call option of the average price type, MC Premium, by applying the Monte Carlo method. We implement 10,000 simulations. MC lower bound and MC upper bound indicate the extremes of the estimate confidence interval in the Monte Carlo approach. The underlying settlement price at maturity is calculated as the average value of electricity prices over one year.

	Strike price	MC Premium	MC lower bound	MC upper bound
In-the-money	59	2.041	1.236	2.846
At-the-money	61	0.198	-0.329	0.725
Out-of-the-money	62	0.003	-0.040	0.046

Before implementing the strategy, we have to estimate the premium $V(0)$ according to scenarios (a), (b) and (c). The premium is obtained by applying the Monte Carlo method[4]. We divide the option lifetime $[0,T]$ of one year into 365 intervals, and we simulate $\Pi = 10,000$ paths of the electricity price according to equation (5.1), discretised according to the Euler scheme (see Kloeden and Platen [49]). For each path, we calculate the average price over the payoff period. Finally, we calculate Π payoffs at maturity and obtain the Monte Carlo price as follows:

$$V(0) = \frac{\sum_{k=1}^{\Pi} \phi^k(T) * e^{-\delta T}}{\Pi}, \qquad (5.8)$$

where $\phi^k(T)$ indicates the payoff at maturity corresponding to the k-th simulation.

We implement the hedging strategy by assuming that the inception is at date 2015/07/06 (that corresponds to the last date of our time series), so that the electricity spot price in the market is 61 €/MWh. Furthermore, we fix the forward price F equal to 61 €/MWh. We contemplate the three scenarios exposed above, (a), (b) and (c), that is when the Asian option is evaluated In-the-money with $K = 59$ €, At-the-money with $K = 61$ € and Out-of-the-money with $K = 62$ €. In addition, we suppose two cases, when the payoff period is the option lifetime, that is one year, and when it is the last month before the maturity.

Case 1: One-year Payoff Period

In this first case, we consider an Asian call option whose underlying settlement price at maturity is the average price of electricity over one year, that is the option lifetime. Thus, if we refer to formula (5.5), n is equal to 365. The results of the Monte Carlo estimation procedure are shown in Table 5.8.

The net profit structure of the strategy at maturity is given by Figure 5.14, which covers the three cases of option valuation.

Looking at Figure 5.14, we can confirm that this strategy works well if there is an expectation of a bear market. We observe that in scenario (a) the company has a low risk profile because it aims at cancelling losses in the event of a bull market, even if it gives up part of its gain in the event of a bear market. Scenario (b) guarantees a good

---

[4]A powerful analytical pricing of Asian-style options is given by Fusai et al. [36].

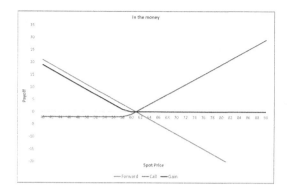

(a)

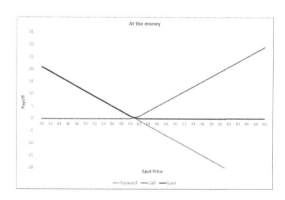

(b)

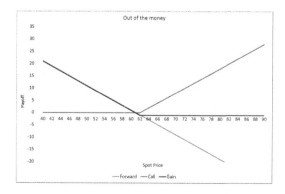

(c)

FIGURE 5.14: Case 1. Net profit strategy at maturity as a function of the electricity spot price at maturity and according to the three scenarios (a), (b) and (c).

TABLE 5.9: Case 2: Results from the Monte Carlo estimation procedure. We estimate the premium of an Asian call option of the average price type, MC Premium, by applying the Monte Carlo method. We implement 10,000 simulations. MC lower bound and MC upper bound indicate the extremes of the estimate confidence interval in the Monte Carlo approach. The underlying settlement price at maturity is calculated as the average value of electricity prices over the last month before maturity.

	Strike price	MC Premium	MC lower bound	MC upper bound
In-the-money	59	1.844	-0.796	4.485
At-the-money	61	0.463	-0.614	1.541
Out-of-the-money	62	0.188	-0.889	1.265

balance between losses and gains. Finally, scenario (c) is reserved to companies that have a higher risk attitude and are ready to come under losses higher than those in the other cases, in order to get a higher gain in the event of a bull market.

Case 2: One-month Payoff Period

In this second case, we consider an Asian call option whose underlying settlement price at maturity is the average price of electricity over the last month before maturity. Thus, if we refer to formula (5.5), n is equal to 30. The results of the Monte Carlo estimation procedure are shown in Table 5.9.

The net profit structure of the strategy at maturity is given by Figure 5.15, which covers the three cases of option valuation.

When we refer to Figure 5.15, we can apply the same considerations done for Case 1 (see Figure 5.14). Moreover, in Figure 5.16 we plot the gains obtained by applying the considered hedging strategy in Case 1 and in Case 2, and according to the three scenarios In-the-money, At-the-money and Out-of-the-money. We find that, with the exception of the scenario In-the-money, it is more convenient to carry out a strategy that comprises the purchase of an Asian call with option period of one year, because the obtained profit at maturity and considering any $S(T)$ is always better with respect to the one in Case 2. Indeed, the curves of Case 1 overlook the ones of Case 2. The judgment overturns for the scenario In-the-money.

5.2   Spark Spread Modelling

The spark spread is an example of commodity spread. It mimics financially the generation costs of electricity for a specific facility. It involves the simultaneous purchase of natural gas futures/forwards and the sale of electricity futures/forwards. The spark spread offers utilities, generators, marketers, and market makers the ability to

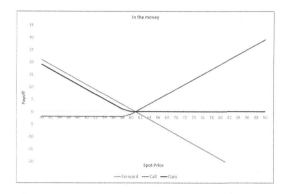

(a)

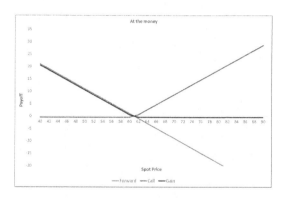

(b)

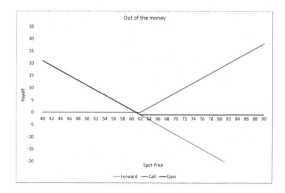

(c)

FIGURE 5.15: Case 2. Net profit strategy at maturity as a function of the electricity spot price at maturity and according to the three scenarios (a), (b) and (c).

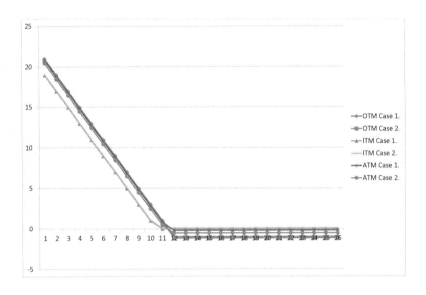

FIGURE 5.16: Case 1 and Case 2 comparison. We plot the gain values of the hedging strategy both in Case 1 and in Case 2. In Case 1, the option period of the Asian call option is one year, whereas in Case 2 it is one month. We indicate with ITM, ATM, OTM, respectively, the three scenarios In-the-money, At-the-money, and Out-of-the-money.

lock in a margin on current and future generation. The classical formula of spark spread is

*Spark Spread = Price of Electricity - [(Price of Gas) * (Heat Rate)] = $/MWh - [($/MMBtu) * (MMBtu / MWh)].*

In order to model the dynamics of the spread we search for the appropriate representation of the electricity price and natural gas price.

5.2.1   The Spot Price Models

Let $(\Omega, F, (F_t)_{t \geq 0}, \mathbb{P})$ be a filtered probability space over an infinite horizon $[0, \infty)$, satisfying the usual conditions. $\mathbb{P}$ is the statistical probability measure. Commodities like electricity have frequently large jumps in the spot price, which in theory should be reflected in the forward prices. We consider a forward on electricity with delivery period $[\tau_1, \tau_2]$ whose price dynamics under the risk-neutral measure $\mathbb{Q}$ are given by the following geometric Brownian motion model with jumps:

$$\frac{dP(t, \tau_1, \tau_2)}{P(t, \tau_1, \tau_2)} = \Sigma^P(t, \tau_1, \tau_2)dW^P(t) + dJ^P(t), \qquad (5.9)$$

where $W^P(t)$ is a Wiener process and $J^P(t)$ defined by

$$dJ^P(t) = (Y_{N^P(t)} - 1)dN^P(t),$$

where $N^P(t)$ follows a homogeneous Poisson process with intensity λ^P, and it is thus distributed like a Poisson distribution with parameter $\lambda^P t$. Y_j is the size of the j-th jump. The Y_j are i.i.d log-normal variables $Y_j \sim N(\mu_Y, \sigma_Y^2)$ that are also independent from the Brownian motion W^P and the basic Poisson process N^P.

Here, $\Sigma^P(t, \tau_1, \tau_2)$ describes the volatility of the forward prices. In order to make the volatility depend on the delivery period explicitly, we choose an average over the delivery period

$$\Sigma^P(t, \tau_1, \tau_2) = \frac{1}{\tau_2 - \tau_1} \int_{\tau_1}^{\tau_2} \sigma^P(t, u)du, \qquad (5.10)$$

where function $\sigma^P(t, u)$ represents the instantaneous volatility. Through function (5.10) we can also give a swap representation of the electricity forward.

According to Clewlow and Strickland [20] we should use the volatility function

$$\sigma(t, u) = \sigma e^{-\alpha(u-t)}, \qquad (5.11)$$

where σ is a constant and α is the rate at which the volatility of increasing maturity forward prices declines and is also the speed of mean reversion of the spot price. On the contrary, as in Kiesel et al. [48], we wish to choose a function that keeps the volatility away from zero as the time to maturity becomes very large. Furthermore, in order to capture the monthly seasonal behaviour of forward prices, we set the function $\sigma^P(t, u)$ depending on the specific month of the year as follows:

$$\sigma^P(t, u) = \sum_{i=1}^{12} (\sigma_2^i + \sigma_1^i e^{-\alpha^i(u-t)})D_i, \qquad (5.12)$$

where dummy variables D_i, $i = 1..., 12$, allow us to estimate the coefficients σ_1^i, σ_2^i and α^i for every month of the year (see Fanelli et al. [32]). Parameter σ_1^i represents the short-term volatility coefficient, σ_2^i the long-term volatility coefficient, and α^i the time decay. Thereby, the volatility function describes the monthly seasonality.

Parameters in function (5.12) can be estimated directly from the prices of options on the spot price of energy or forward contracts or, alternatively, by best fitting to historical volatilities of forward prices.

Then

$$\Sigma^P(t, \tau_1, \tau_2) = \sum_{i=1}^{12} \left(\sigma_2^i + \frac{\sigma_1^i(e^{-\alpha_i(\tau_1-t)} - e^{-\alpha_i(\tau_2-t)})}{\alpha_i(\tau_2 - \tau_1)} \right) D_i. \qquad (5.13)$$

We propose the model (5.9) also for the forward on natural gas with delivery period $[\tau_1, \tau_2]$, so that

$$\frac{dF(t,\tau_1,\tau_2)}{F(t,\tau_1,\tau_2)} = \Sigma^F(t,\tau_1,\tau_2)dW(t) + dJ(t), \qquad (5.14)$$

where $W^F(t)$ is a Wiener process and $J^F(t)$ defined by

$$dJ^F(t) = (Y_{N^F(t)} - 1)dN^F(t),$$

where $N^F(t)$ follows a homogeneous Poisson process with intensity λ^F, and it is thus distributed like a Poisson distribution with parameter $\lambda^F t$. Y_j is the size of the j-th jump. The Y_j are i.i.d log-normal variables $Y_j \sim N(\mu_Y, \sigma_Y^2)$ that are also independent from the Brownian motion W^F and the basic Poisson process N^F.

Again $\Sigma^F(t,\tau_1,\tau_2)$ describes the volatility of the forward prices. In order to make the volatility depend on the delivery period explicitly, we choose an average over the delivery period

$$\Sigma^F(t,\tau_1,\tau_2) = \frac{1}{\tau_2 - \tau_1} \int_{\tau_1}^{\tau_2} \sigma^F(t,u)du, \qquad (5.15)$$

where function $\sigma^F(t,u)$ represents the instantaneous volatility.

The monthly seasonal function $\sigma^F(t,u)$ is:

$$\sigma^F(t,u) = \sum_{i=1}^{12} (\beta_2^i + \beta_1^i e^{-\delta^i(u-t)})D_i, \qquad (5.16)$$

where dummy variables D_i, $i = 1...,12$, allow us to estimate the coefficients β_1^i, β_2^i and δ^i for every month of the year (see Fanelli et al. [32]). $\Sigma^F(t,\tau_1,\tau_2)$ is obtained as in (5.13).

5.2.2 Data Analysis

The data set consists of daily calendar forward prices of the Italian electricity and natural gas, spanning from 2 January 2014 to 1 February 2016, resulting in 470 price quotes for each commodity. The two time series are plotted in Figure 5.17. In Figures 5.18 and 5.19 we show the histograms of logarithmic returns.

We first check for outliers in the considered time series. We analyse the time series of the logarithmic forward prices and we follow this procedure: given the lower and the upper quartiles, respectively Q1 and Q3, an observation is called an outlier if it is smaller than Q1-3·(Q3-Q1), or larger than Q3+3·(Q3-Q1). We found no outliers.

In order to eliminate the trend and the seasonality from the price time series, we model the logarithmic forward prices with the following mean level function:

$$\log X(t,\tau_1,\tau_2) = a_0 + a_1 t + a_2 \cos(2\pi t). \qquad (5.17)$$

We assume 252 trading days in a year, and function (5.17) consists of a linear trend and a seasonal term explaining possible variations over the year. We estimate the

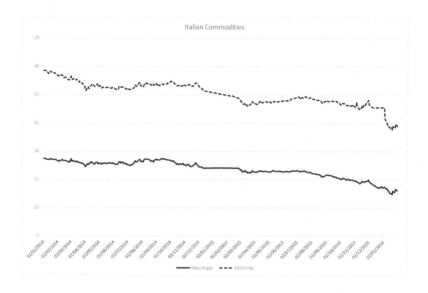

FIGURE 5.17: Italian commodity forward prices

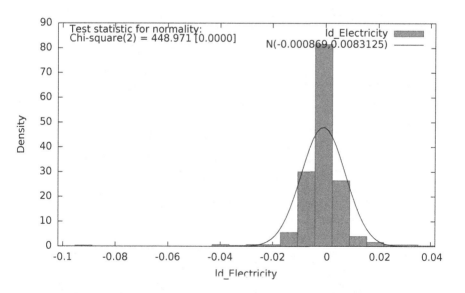

FIGURE 5.18: Histogram of daily changes in the logarithmic electricity forward prices

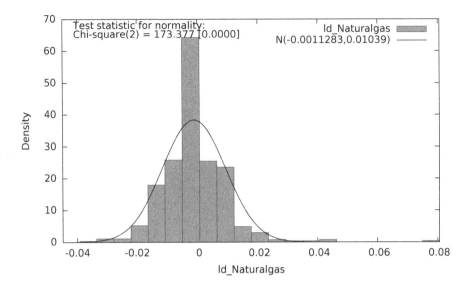

FIGURE 5.19: Histogram of daily changes in the logarithmic natural gas forward prices

parameters of function (5.17) by applying the least squares approach, both for the electricity prices and for the natural gas prices. Results are displayed in Tables 5.10 and 5.11.

After estimating the mean function, we remove the effect of the seasonality by subtracting the values obtained by using function (5.17) from the actual logarithmic prices. The detrended and deseasonalized logarithmic prices are plotted in Figure 5.20.

The histograms of the logarithmic returns are presented in Figures 5.21 and 5.22. For electricity the skewness is -2.7683, the kurtosis 33.810. For natural gas, the skewness is 1.3118, the kurtosis 8.7327. In Figure 5.23 we plot the detrended and deseasonalized returns.

Observe that returns in Figure 5.23 have mostly small fluctuations around zero, but from time to time rather extreme jumps appear. This motivates the use of a mix of a geometric Brownian motion and a jump process in (5.9) and (5.14). We apply recursive filtering procedures to identify jumps and estimate jump process parameters (see Clewlow and Strickland [21]). The procedure identifies as a jump the return which deviates in absolute value more than a fixed level from the mean. The most commonly used levels are two or three standard deviations. The filtering is performed recursively in the sense that after identifying jumps, these are removed and the level is recalculated for a new round of identification of jumps. We iterate until the level is unchanged, and no new jumps are found in the procedure. We use a level of three standard deviations and the results are presented in Tables 5.12 and 5.13.

TABLE 5.10: Fitted parameters of function (5.17) for electricity

Model: OLS, using observations 1:1–94:5 ($T = 470$)
Dependent variable: l_Electricity

	Coefficient	Std. Error	t-ratio	p-value
const	4.04829	0.00365776	1106.7655	0.0000
time	−0.144482	0.00342332	−42.2051	0.0000
time2	−0.0134320	0.00267682	−5.0179	0.0000

Mean dependent var	3.914128	S.D. dependent var	0.086321
Sum squared resid	0.724445	S.E. of regression	0.039386
R^2	0.792702	Adjusted R^2	0.791814
$F(2,467)$	892.8965	P-value(F)	2.7e–160
Log-likelihood	854.7432	Akaike criterion	−1703.486
Schwarz criterion	−1691.028	Hannan–Quinn	−1698.585
$\hat{\rho}$	0.984714	Durbin–Watson	0.044583

TABLE 5.11: Fitted parameters of function (5.17) for natural gas

Model: OLS, using observations 1:1–94:5 ($T = 470$)
Dependent variable: l_Naturalgas

	Coefficient	Std. Error	t-ratio	p-value
const	3.36614	0.00560440	600.6256	0.0000
time	−0.231758	0.00524518	−44.1850	0.0000
time2	−0.0196854	0.00410140	−4.7997	0.0000

Mean dependent var	3.150827	S.D. dependent var	0.137266
Sum squared resid	1.700717	S.E. of regression	0.060347
R^2	0.807544	Adjusted R^2	0.806720
$F(2,467)$	979.7640	P-value(F)	7.8e–168
Log-likelihood	654.1943	Akaike criterion	−1302.389
Schwarz criterion	−1289.930	Hannan–Quinn	−1297.487
$\hat{\rho}$	0.990219	Durbin–Watson	0.029624

In order to keep into consideration the fact that we can have both positive and negative jumps, we assume that there are two different Poisson processes for each commodity. The intensity values are reported in Table 5.14.

Finally, in Figures 5.24, 5.25, 5.26 and 5.27 we plot the autocorrelation function, the partial autocorrelation function and the QQ test for the filtered log-returns of electricity and natural gas that confirm the hypothesis of a geometric Brownian motion.

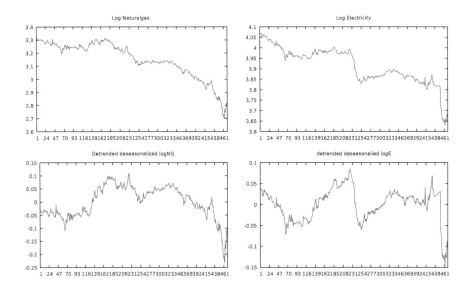

FIGURE 5.20: Detrended and deseasonalized logarithm of commodity forward prices

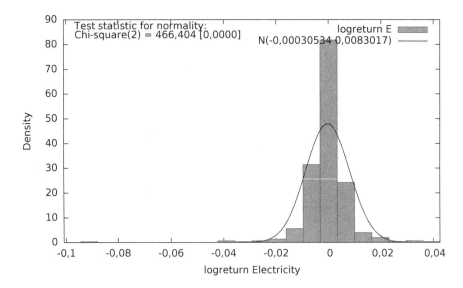

FIGURE 5.21: Histogram of daily changes in the logarithmic electricity forward prices after trend and seasonal components were eliminated

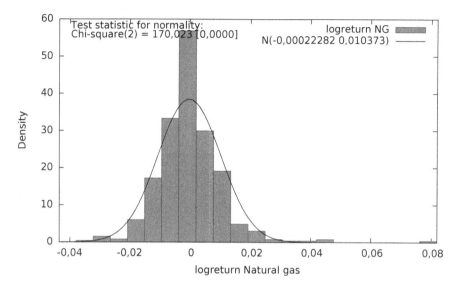

FIGURE 5.22: Histogram of daily changes in the logarithmic natural gas forward prices after trend and seasonal components were eliminated

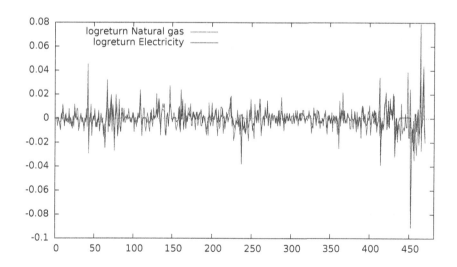

FIGURE 5.23: Logarithmic commodity price returns after trend and seasonal components were eliminated

TABLE 5.12: Electricity: Parameter estimates from jump-recursive filter estimation procedure

Iteration	Return SD	Jump Threshold	No. Jumps	Jump Intensity	Mean Jump	Jump SD
0	0.0083	0.0249	-	-	-	-
1	0.0058	0.0175	8	0.0170	-0.0151	0.0461
2	0.0052	0.0155	16	0.0340	-0.0027	0.0366
3	0.0049	0.0147	21	0.0447	-0.0028	0.0327
4	0.0047	0.0141	25	0.0532	-0.0012	0.0306
5	0.0046	0.0138	27	0.0574	-0.0021	0.0296
6	0.0046	0.0137	28	0.0596	-0.0016	0.0292
7	0.0046	0.0137	28	0.0596	-0.0016	0.0292

TABLE 5.13: Natural gas: Parameter estimates from jump-recursive filter estimation procedure

Iteration	Return SD	Jump Threshold	No. Jumps	Jump Intensity	Mean Jump	Jump SD
0	0.0104	0.0311	-	-	-	-
1	0.0088	0.0265	6	0.0128	0.0341	0.0375
2	0.0083	0.0249	12	0.0255	0.0121	0.0393
3	0.0082	0.0247	13	0.0277	0.0092	0.0390
4	0.0082	0.0247	13	0.0277	0.0092	0.0390

TABLE 5.14: Jump intensities estimates

Commodity	λ^i_+	λ^i_-
Electricity $i = E$	0.0319	0.0287
Natural gas $i = NG$	0.0149	0,0128

5.3   Arbitrage Strategy in Commodity Markets

Fanelli [29] reviews and discusses some arbitrage strategies in several commodity markets. However, most arbitrage model theories have been developed for asset (equity) markets. Developing this statistical arbitrage trading model, we apply these methodologies to the commodities futures trading market.

In order to exploit arbitrage opportunities arising on financial markets investors and speculators have been interested in studying and developing quantitative methods and trading strategies. The first trading strategy used in the 1980s was called "Pairs Trading", where short-term mispricings in a pair of similar securities were exploited through arbitrage trading. Traders using this strategy identified trading opportunities by using graphic analysis of the trend and reversion trend that exist between the prices dynamic of two assets that have similar characteristics. The asset pairs were

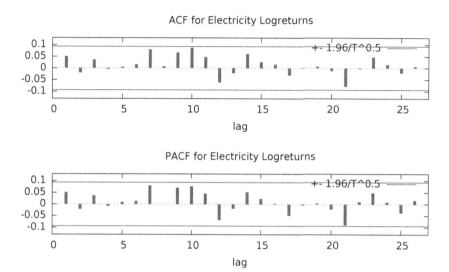

FIGURE 5.24: ACF and PCF of the filtered electricity logreturns

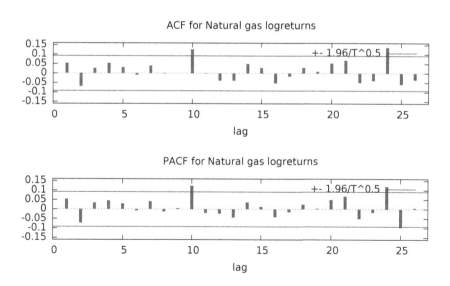

FIGURE 5.25: ACF and PCF of the filtered natural gas logreturns

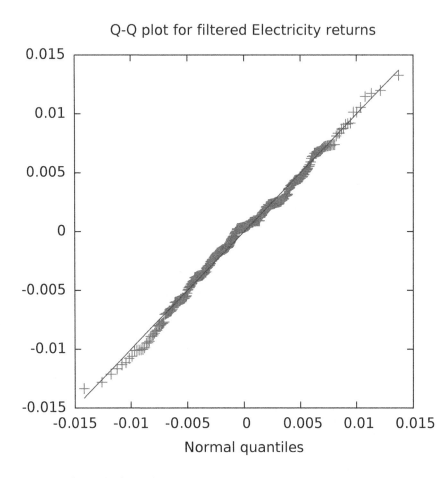

FIGURE 5.26: QQ test for filtered electricity forward returns

selected on the basis of intuition, economic fundamentals, long-term correlations or simply on past experience.

By 2000 the growing demand for models that could correctly describe more sophisticated trading strategies, led to the development of so-called statistical arbitrage strategies. Statistical arbitrage (StatArb) can best be described as the attempt to profit from pricing inefficiencies which are identified using mathematical and statistical models. The term StatArb was used for the first time in the 1990s and it was widely used by operators in financial markets until 2002. In 2000 dramatic changes in market dynamics led to weak performance of the models and, consequently to a loss of confidence in StatArb methods. According to Pole [57], more accurate and advanced algorithms aroused a new interest in statistical arbitrage models in 2006.

According to Burgess [12] and Bondarenko [10] a StatArb is a generalization of the traditional zero-risk or pure arbitrage. In the latter case, gains are received with no

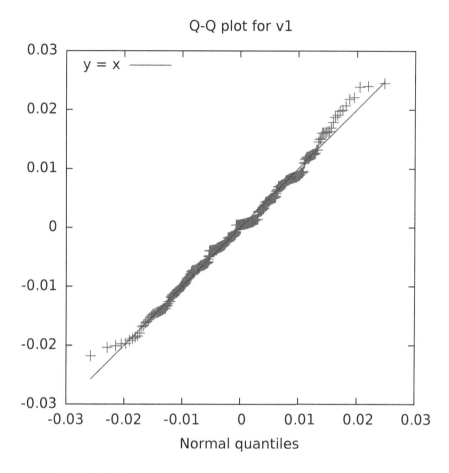

FIGURE 5.27: QQ test for filtered natural gas forward returns

possibility of losses. Fair-price relationships between asset pairs with identical cash-flows are constructed and pure arbitrage opportunities, which are identified when prices deviate from these relationships.

Instead, we refer to a StatArb when strategy's payoff can be negative in some elementary states, as long as the average payoff in each final state is no-negative. As a general rule, StatArb stems from the awareness that, on market, zero-risk arbitrage opportunities don't exist due to several uncertainty factors; such as the fluctuations in difference between spot and futures prices, prior to the expiry date, that cause the so-called basis risk. Basis risk, on one hand, is a source of uncertainty when positions on securities have to be marked to market at current prices by operators due to exchange regulations and companies' internal requirements. On the other hand, it is a source of opportunity because an arbitrageur can assume a position on a security and revert the trade before the expiry date when he or she realizes profits.

Many authors, like Burgess [13], Elliott, Van Der Hoek, and Malcolm [26], Do, Faff, and Hamza [25] and Bertram [6], proposed quantitative methods to exploit StatArb opportunities on stock markets. Usually, a methodology for StatArb consists of three steps:

1. synthetic assets are constructed as combinations of other existing assets to replicate the value of a target asset or target portfolio, so that a statistical fair-price relationship is obtained. The assets are selected in such a way that the deviations from the fair-price relationship, called statistical mispricings, have potentially predictable components in their dynamics. The mispricings have to exhibit a strong mean reversion;

2. statistical arbitrage opportunities are identified through statistical estimates and forecasts of the changes in mispricing dynamics;

3. appropriate trading rules are implemented to profit.

We consider a set of commodities, crude oils, select a target commodity and replicate its price with a portfolio consisting of other commodities, such that statistical mispricing time series are constructed. This first step is developed upon the econometric concept of cointegration. In a second step, we have verified that the mispricing dynamics fulfil the predictability conditions. We employed standard tests for autoregressive, mean-reverting and non random walks dynamics; we have referred to Burgess [12] applying variance ratio tests to verify the mean-reverting behaviour of the mispricing dynamics. The third step of the methodology consists in implementing the so-called implicit statistical arbitrage strategies. These strategies exploit directly the typically mean-reverting nature of the mispricing dynamics, without the development of an explicit forecasting model.

Finally we build and optimize a forecasting implicit statistical arbitrage model on in-sample data which consisted of one crude oil and some distillates. We then have tested the model on out-of-sample data considering performance indicators.

5.3.1 Mispricing Investigation

In this section we describe the first and second steps of the methodology adopted for building a statistical arbitrage model. In the first step we construct the so-called statistical mispricing time series following the cointegration approach described by Burgess [12]. We individuate a target commodity and select other commodities whose combination forms a replication portfolio of target commodity. The deviation between the target commodity price and the replication portfolio value represents a statistical mispricing. The use of a prices' combination of prices instead of a single price has a financial meaning. In fact, combinations (relative value) are statistically independent of market-wide risks and influenced by commodity specific aspects. This statement is the core of some traditional asset pricing models such as CAPM (Capital Asset Pricing Model) and APT (Arbitrage Pricing Theory). The point is that the noises or stochastic components are common to many assets, so an appropriate asset combination would be immune to market-wide risks. Consequently, replication portfolio dynamics are affected only by asset specific component dynamics that are potentially more predictable.

In theory, we consider a universe of commodities and a particular target commodity, given prices at some point in time. A synthetic asset can then be constructed as a linear combination of other similar commodities, such that the following statistical fair-price relationship holds for some time $t \geq 0$

$$E[P(t)|\mathbf{Z}(t)] = X(t), \tag{5.18}$$

representing a long-term relationship among variables, where $P(t)$ denotes the target asset price, $\mathbf{Z}(t)$ denotes the commodity prices' vector, $X(t)$ denotes the synthetic asset prices and $E[\cdot]$ is the expectation. It follows from equation (5.18) that the statistical mispricing at time t is given by the deviation from the fair-price relationship

$$M(t) = P(t) - X(t). \tag{5.19}$$

The mispricing dynamics contain predictable components that can be exploited on the basis of statistical arbitrage trading strategies. The methodology most commonly adopted to construct the series of the statistical mispricings is based on cointegration techniques. A cointegration regression is used to obtain the linear combination coefficients of commodity prices which are more correlated with the target price, such that they share the same common trend, long-run equilibrium. The linear combination coefficients, constituting the vector of cointegration, are estimated by regressing a set of historical prices $Z^i(t)$ of the n commodities constituting the replication portfolio, such that

$$X(t) = \sum_i \omega_i Z^i(t), \qquad i = 1, ...n, \tag{5.20}$$

where ω_i =respective weights. Substituting (5.20) back into (5.19) we obtain the corresponding statistical mispricings

$$M(t) = P(t) - \sum_i \omega_i Z^i(t), \qquad i = 1, ...n, \tag{5.21}$$

which are the values of a portfolio of commodities $\{P(t), Z^1(t), Z^2(t), ..., Z^n(t)\}$ associated with the corresponding weights $\{1, -\omega_1, -\omega_2, ..., -\omega_n\}$. In other words, $M(t)$ represents the difference between the target price and the value of the replication portfolio and can be seen as a "stochastically de-trended" version of the original target price with respect to the observed time series. Hence, $M(t)$ behaves as a proxy of non-directly observed risk factors, which determine a stochastic trend that is common to many market prices. With this procedure we create a synthetic asset having an exposure of the underlying risk factors (not necessarily directly observable) similar to that driving the target commodity dynamics.

In practice, we consider a data set formed by a series of three crude oils spanning from 25/10/2000 to 19/10/2009, all prices quoted weekly in US dollars per barrel (bbl). We use the WTI future first month traded at the New York Mercantile Exchange

TABLE 5.15: Cointegration regression for crude oils

Variable	Coefficient	Std. Error	t-Statistic	Prob.
constant	1.617633	0.194177	8.330716	0.0000
BRENT	1.193785	0.0406112	9.39531	0.0000
DUBAI	-0.217020	0.042228	-5.139301	0.0000

R-squared	0.995153
Adjusted R-squared	0.995131
S.E. of regression	1.832739
Akaike info criterion	4.055987
Schwarz criterion	4.082885
F-statistic	47012.19
Prob(F-stat)	0.000000
Durbin-Watson statistic	0.256805

(Nymex), the Brent future first month traded at the International Petroleum Exchange in London (IPE) and the Dubai forward first month quote by Platt's.

We refer to WTI future as the target commodity and Brent future and Dubai forward represent the constituents of the replication portfolio. Then, we carry out the cointegration analysis to find coefficients of the commodities' linear combination that replicates the WTI future. The results of the cointegration regression on 461 observations are shown in Table 5.15. The test statistic of the Augmented Dickey-Fuller (ADF) test for residuals is -3.75029 with a p-value of 0.04869, so that a cointegration relationship is evidenced. Statistical mispricing time series are plotted in Figure 5.28.

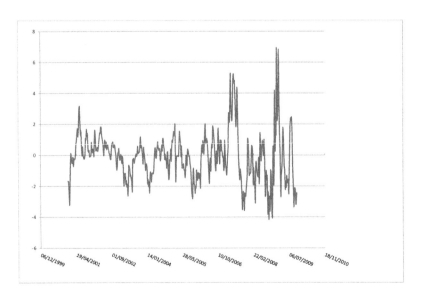

FIGURE 5.28: Crude oil statistical mispricing

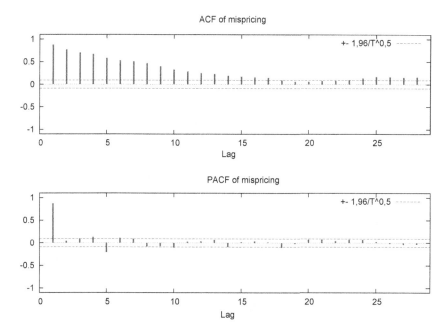

FIGURE 5.29: Autocorrelation function

Now, we investigate if there exists a predictability condition in our model, namely we have to identify potentially predictable components in the dynamics of the statistical mispricing time series. The autocorrelation function of the time series is used to examine the short-term effects. As we can see from Figure 5.29, autocorrelation coefficients with a value different from zero indicate that the future value of time series is related to the past value and hence the presence of a predictable component is expected.

We look at the above results for unit root tests, Table 5.15, to verify the stationarity of the time series. The stationarity is asserted by the value -3.75 of the ADF statistic test, even if the acceptable but high value of the p-value 0.04869 could mean an absence of mean reversion. A theoretical problem about the low power of classical statistical tests, like DF and ADF, to clearly identify the prediction conditions is well known in the econometrics field. In this context, we use the more robust test of variance ratio to verify if the dynamics of the time series deviate from the random walk behaviour. The analysis of the variance ratio functions of the time series allows us to find out a mean-reverting nature. The variance ratio function is defined as the normalized ratio long term variance, calculated over a period τ, to a single period variance. If values of variance ratio are bigger than one, then the long-term variance is higher than the short-term one and the variance ratio function is increasing. The net effect over timescale τ is of positive autocorrelation or trending behaviour. On the contrary, if the variance ratio function is decreasing because the variance ratios are smaller than

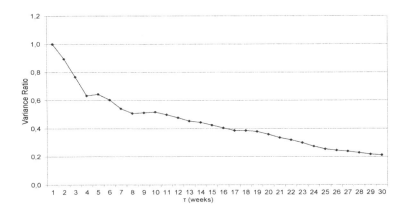

FIGURE 5.30: Variance ratio function

one, for any τ, and so the long-term variance is lower than the short-term one, the net effect over the period is of negative autocorrelation or mean-reverting behaviour. Figure 5.30 displays the variance ratio function of mispricing relative to our analysis. Mispricing dynamics follow a mean-reverting process, asserting the sure evidence of predictable components.

5.3.2 Statistical Arbitrage Trading Strategies

In this section we discuss the third step of our methodology. We aim at investigating suitable trading rules that identify trading signals for opening and closing positions on the market. We review the implicit statistical arbitrage strategies described in Burgess [12] and test them on commodity price data.

Implicit statistical arbitrage (ISA) trading strategies are based on trading rules that rely implicitly on the mean-reverting behaviour of the mispricing time series. If we consider a mispricing model like the one described in the previous section, the fundamental requirement for the application of the ISA trading rules is that the mispricing dynamics mean-revert. In fact, if in the long-term the misprisings reduce as prices change, an operator, who has previously opened a position, can realize profits provided that the impact of transaction costs, supported to employ the strategy, turns out less than the impact of the mispricing gain component.

ISA trading rules define the sign and the magnitude of the mispricing portfolio components, that is the target commodity and the replication portfolio. We implement the following trading rules.

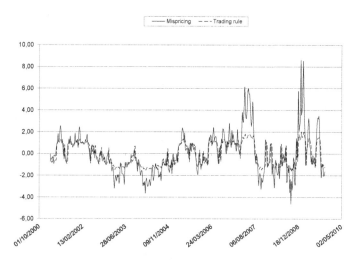

FIGURE 5.31: Trading rule example: the dashed line represents the trading rule in equation (5.22) with parameter $K = 0.33$; it gives signals for trading the mispricing given by the solid line

Rule 1

Given a time interval $[t - j, t]$, the basic trading rule is

$$S(t)^k = -sign(M(t-1))|M(t-1)|^k. \qquad (5.22)$$

The mispricing portfolio must be sold when $S(t)^k$ is negative and bought when it is positive. The holding magnitude varies as a function of the size of the previous mispricing through the sensitivity parameter. An example of a trading rule as a function of the time is displayed in Figure 5.31.

In order to reduce the investor risk attitude in mispricing trading, we use the moving-average strategy $S^h(t)$, defined as follows:

$$S^h(t) = \frac{1}{h} \sum_{j=1}^{h} S(t-j)^k, \qquad (5.23)$$

where $h > 0$ is the moving average parameter. Furthermore, any transaction, made according to any trading strategy, implies some costs and, obviously, every operator wants to optimize the tradeoff between costs and gains from the exploitation of trading signals. In order to do this we consider the following smooth strategy:

$$S^O(t) = (1-O)S_t^k + OS^O(t-1), \qquad (5.24)$$

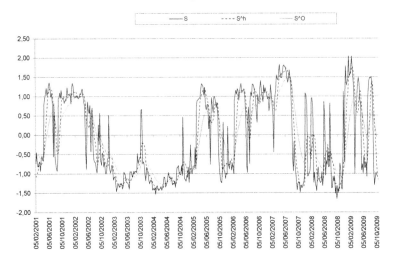

FIGURE 5.32: Trading signal comparison: the solid line represents the dynamics of the trading rule in equation (5.22) with parameter $K = 0.33$; the dashed line describes the dynamics of the trading rule in equation (5.23) with parameters $K = 0.33$, and $h = 6$; the dotted line represents the dynamics of the trading rule in equation (5.24) with parameters $K = 0.33$, $h = 6$ and $O = 0.9$

where $O \geq 0$ is a smoothing parameter. By increasing the values of h and O, on one hand the number of transactions comes down so that the transaction costs decrease; on the other hand, the accuracy of the smoothed trading signal diminishes. This behaviour is illustrated in Figure 5.32 where we compare the three trading strategies (5.22), (5.23) and (5.24) as functions of the time, assuming $K = 0.33$, $h = 6$ and $O = 0.9$. From the figure we deduce that strategy function (5.22) changes more frequently with respect to function (5.23) and has larger amplitude compared with function (5.24). Function (5.24) has a very smooth trend in comparison with the other functions meaning that, ceteris paribus, adopting this strategy means the number of transactions reduces.

We estimate the following performance indicators to evaluate the performance of the trading rules. The first indicator $R(t)$ is the mark-to-market profit & loss, which evaluates the return gotten over a generic trading time period $[t-1, t]$ by applying any trading rule. Let us consider for example rule (5.22), the mark-to-market profit & loss at time t is computed by the following formula

$$R(t) = S(t)^k \frac{\Delta M(t)}{S(t) + V(t)^h} - c|\Delta S(t)^k|, \tag{5.25}$$

where $\Delta M(t) = M(t) - M(t-1)$, $\Delta S(t)^k = S(t)^k - S(t-1)^k$, c is the percentage transaction costs and $S(t) + V(t)^h$ is the sum of the mispricing portfolio components. Then, we can substitute $S(t)^k$ with the other trading strategies (5.23) and (5.24) in order to obtain the return of those strategies.

A strategy profitability indicator is the cumulative mark-to-market profit & loss, $\rho(t)$, that represents the total return or cumulative profit of a strategy from the inception $t = 0$ to the generic trading date t. It is computed as the cumulative sum of the $R(s)$, $s = 0, ..., t$:

$$\rho(t) = \sum_{s=0}^{t} R(s). \qquad (5.26)$$

We can use the indicator $\rho(t)$ to compare the performances of strategy (5.22) according to different values of k.

A performance indicator that takes into account not only the level of profit, but also the level of strategy risk, measured by the variability of profits, is the Sharpe Ratio. The Sharpe Ratio calculated at date t is $\Pi(t)$. As in the traditional sense, it measures the profit per unit of risk. In this context of the statistical arbitrage, it is calculated as the ratio between the annualized mean profitability of the strategy and its annualized standard deviation of the profits:

$$\Pi(t) = \frac{\frac{1}{t}\sum_{s=1}^{t} R(s)}{\sqrt{\frac{1}{t}\sum_{s=1}^{t} \left[\left(R(s) - \frac{1}{t}\sum_{s=1}^{t} R(s) \right) \right]^2}}. \qquad (5.27)$$

Figure 5.33 shows the cumulative profit functions for $k = 0, 0.5, 1$ and confirms the value $k = 1$ ensures the greater profit.

Using the performance indicators described above, we compare the trading strategies according to certain parameters and we investigate the most efficient one. We

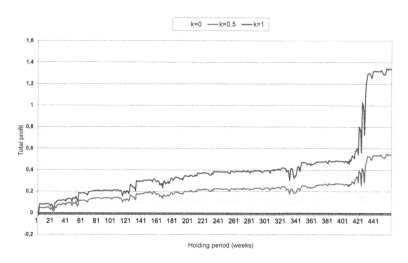

FIGURE 5.33: Total return function comparison: three cumulative profit functions given by formula (5.26) are shown according to different values of parameter k of the strategy in equation (5.22)

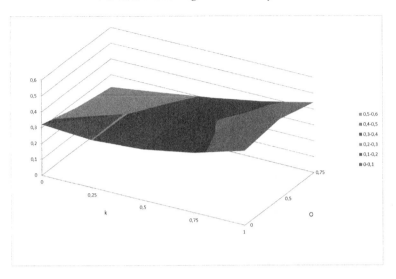

FIGURE 5.34: Optimal trading strategy: the surface represents the total return, given by formula (5.26), of the strategy in equation (5.24); the parameter k varies between 0 and 1 and the parameter O varies between 0 and 0.75; transaction cost percentage is 0.25%

apply the strategy $S^O(t)$, equation (5.24), on our mispricing data letting k vary between 0 and 1 and O vary between 0 and 0.75. We calculate the values of $\rho(t)$ at the last observation for each k and represent them in the Figure 5.34. From the figure we can deduce that the optimal strategy ensuring the maximum profit is when $k = 1$ and $O = 0.5$ and assuming a cost percentage equal to 0.25%. We use this optimal rule to test the effectiveness of our statistical arbitrage strategy by an out-of-sample analysis as described in the following subsection.

5.3.3   Forecasting Model

In this section we aim at building a forecasting model that, by using ISA strategies, exploits trading signals and produces profits[5].

The considered data set is formed by the Brent future first month traded at the International Petroleum Exchange in London (IPE) and a series of five products spanning from 25/10/2000 to 19/10/2009, all prices quoted weekly in US dollars per barrel (bbl). Regarding the distillate, we use the "Rotterdam Premium Gasoline unleaded Fob Cargoes" (PUR) quoted by Platt's as a light distillate, the Gas Oil future first month traded in IPE as a middle distillate product and finally the North West Europe (NWE) "Low Sulphur Fuel Oil Cargoes"(LSFO), the NWE "High Sulphur Fuel Oil Barger" (HSFO) and the NWE "Cts 180 Bunker Fuel Oil" as heavy distillates

---

[5]We suggest to refer to Cerqueti et al. [15] and Cerqueti and Fanelli [14] for an additional study on the long-term analysis and predictability of crude oil portfolios.

TABLE 5.16: Cointegration regression for Brent futures and distillates

Variable	Coefficient	Std. Error	t-Statistic	Prob.
LSFO	-0,218887231	0,05397707	-4,055189171	6,64605E-05
HSFO	0,568914197	0,138187861	4,11696219	5,17705E-05
Cst180	-0,129489633	0,116300872	-1,113402081	0,266576047
PUR	0,20284117	0,020861772	9,723103653	3,13428E-19
Gas Oil	0,514229011	0,018819257	27,32461759	5,42177E-78

products. We consider a portfolio characterised by the virtual refinery model, based on the technology relation between the crude, that is the input for the refinery, and the products that represent the output. Using a large sample of data, we can split the data into two parts: data from 25/10/2000 to 26/12/2005 are used for developing the model through an in-sample analysis; data from 02/01/2006 to 19/10/2009 are used for testing the model and evaluating its out-of-sample performance.

According to the methodology described in the previous sections, the coefficients of the linear combination of distillates forming the replication portfolio are calculated on in-sample data by a cointegration regression, referring to Brent as dependent variable. The results of the regression are in Table 5.16.

In Figure 5.35 we display the variance ratio profile for the mispricings. The deviation of the mispricing process from a random walk is strong and the mispricing clearly follows a mean-reverting process, evidencing the presence of predictability components.

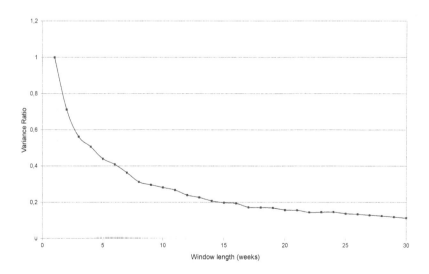

FIGURE 5.35: Variance ratio function

TABLE 5.17: In-sample performance of the basic strategy (5.22)

Year	2001	2002	2003	2004	2005
Total Return	12.79%	6.43%	6.52%	10.63%	21.94%
Sharpe Ratio	2.01	2.03	1.95	1.90	0.03
Profitable weeks	58.33%	57.69%	59.62%	61.54%	57.69%

TABLE 5.18: Out-of-sample performance of the basic strategy (5.22)

Year	2006	2007	2008	2009
Total Return	19.29%	57.07%	137.14%	53.56%
Sharpe Ratio	2.07	2.03	2.51	2.86
Profitable weeks	45.65%	66.04%	70.59%	66.67%

We now apply the described basic strategy both to in-sample and to out-of-sample data and compare the performances to assess the functioning of the model in a forecasting perspective. We use trading strategy (5.22) with $k = 1$, and measure the trading performance for each year using the total return, the annual Sharpe Ratio and calculating the percentage profitable weeks as the percentage of periods corresponding to positive values of return (5.25). The results for in-sample data are illustrated in Table 5.17, while those for out-of-sample data are in Table 5.18.

From this simple analysis we find out that our forecasting model performs well for the out-of-sample years, in line with the results obtained on in-sample data. This statement is confirmed by Figure 5.36, which shows the paths over the entire reference period 2001-2009 of the actual Brent future prices and of the forecast prices

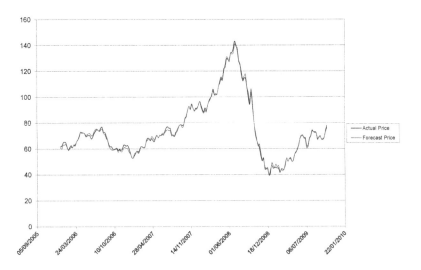

FIGURE 5.36: Brent future actual prices vs. forecast prices

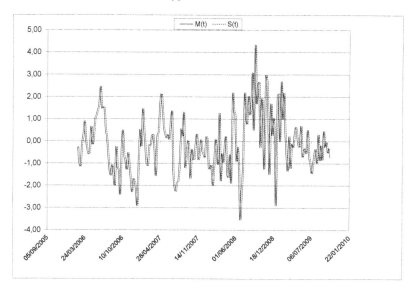

FIGURE 5.37: Brent's and distillates' trading strategy

calculated using the estimated regression parameters (contained in Table 5.16). Furthermore, from observing Figure 5.36 we can assert that the fair-price relationship built by in-sample analysis holds over time.

In order to illustrate the effectiveness of the implemented strategy, we consider the results for years 2008 and 2009 displayed in Table 5.18. We expect a total return of 137.14% for year 2008 greater than the one of 53.56%, expected for year 2009. This profitability scenario is confirmed by the presence of a higher percentage of profitable days for 2008 with respect to 2009. We come to opposite conclusions if we consider the index of the Sharpe Ratio as an indicator of performance because it is lower for 2008 than for 2009. In fact, the amplitude and oscillations' frequency of the graph of our strategy (Figure 5.37) reflect a higher variability of profits due to the volatility of the Brent future price that is greater in 2008 than in 2009. Consequently, an optimal strategy can be developed and updated daily taking into consideration the model forecasts and the expected values of the three indices of performance (total return, Sharpe Ratio and profitable periods), so that any trading decision will be taken in line with the specific risk profile.

Chapter 6

Essential Statistics and Data Analysis

6.1   Plotting Time Series

When we aim at understanding the behaviour of a commodity price and the best representation of it, we start by looking at the plot of the price time series. For example, we can observe the futures price plot and use the cost-of-carry relationship to have insights into the price dynamics. According to the cost-of-carry relationship seen in Chapter 3, the futures price, $F(t,T)$, of a commodity must be equal to the cost of acquiring the physical commodity at price $S(t)$ and carrying it until the future maturity T:

$$F(t,T) = S(t)e^{[r(t)+m(t)-c(t)](T-t)},$$

where $r(t)$ is the riskless interest rate on the date t, $m(t)$ is the storage cost (per unit of time and per dollar's worth of commodity) and $c(t)$ is the convenience yield, all continuously compounded. If we define $\gamma(t) = c(t) - m(t)$ we obtain

$$F(t,T) = S(t)e^{[r(t)-\gamma(t)](T-t)}.$$

In Figures 6.1, 6.2, 6.3 and 6.4 we plot the futures price time series of the WTI crude oil, natural gas, coffee and gold.

In Figure 6.1 we observe that the price is expected to rise until December 2012 and then to decrease gradually. From the cost-of-carry relationship we infer that strong supply and negative cost-of-carry discourage additional storage.

For the case of natural gas, Figure 6.2, we say that the gas price is expected to be high in winter because it is scarce, and to be low in summer because of strong supply. Therefore, a surplus in the physical market and high inventories drive down the convenience yield so that this fact discourages further storage. Furthermore, we clearly observe a seasonal behaviour.

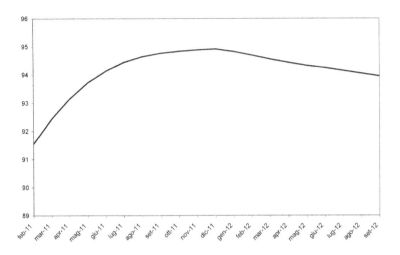

FIGURE 6.1: WTI futures price time series

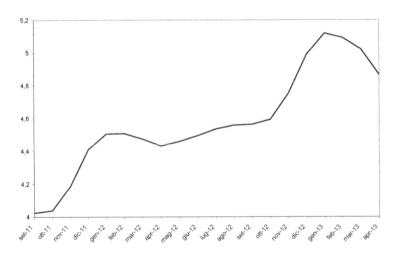

FIGURE 6.2: Natural gas futures price time series

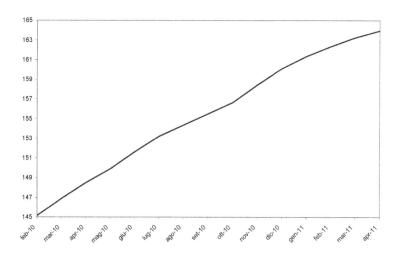

FIGURE 6.3: Coffee futures price time series

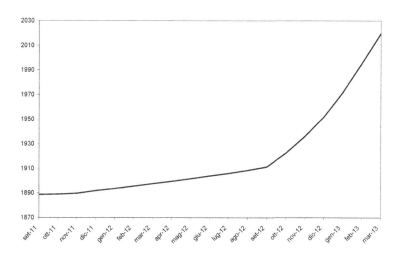

FIGURE 6.4: Gold futures price time series

From Figure 6.3 we deduce that the gold price should increase and the strong supply of coffee generates a negative cost-of-carry.

Finally, also the gold price is expected to rise by the cost-of-carry, due to the increase of risk-free rate and storage costs especially from September 2012; see Figure 6.4.

6.2   Probability Distribution Analysis

In Table 6.1 we compare the main characteristics of two data analysis approaches: time series analysis and probability distribution analysis. A price distribution defines the probabilities of prices taking on various values. If we are analysing actual data, the distribution is defined by the "path" of prices observed over the time period. If we are simulating a model, the distribution shows all the possible values that the price might take on over some time period with associated probabilities. The n-th moment of a random variable X distribution is the expected value of the random variable raised to the n-th power. Depending on the nature of the random variable, we can calculate the n-th moments in two ways:

$$E[X^n] = \int_{-\infty}^{+\infty} f(x)x^n \quad \text{for continuous random variables,}$$

or

$$E[X^n] = \sum_{i=1}^{n} p(x_i)x_i^n \quad \text{for discrete random variables,}$$

where

- $E[\cdot]$ represents the expected value,

- x is the realisation of the random variable X,

TABLE 6.1: Time series analysis vs. probability distribution analysis

	Time series analysis	Probability distribution analysis
Purpose	Analyses price changes from day to day	Analyses price behaviour over a time period
Good for	Parameter calibration and seasonal calibration	Examining, benchmarking, and selecting models, getting insights about option models
Use in business	Relatively common	Uncommon, but sometimes more effective

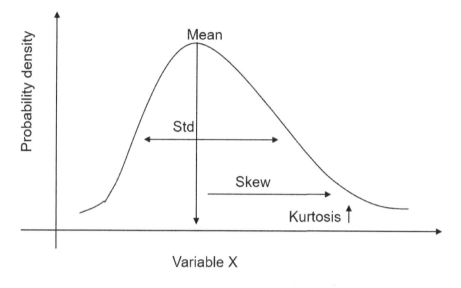

FIGURE 6.5: Moments of a probability distribution

- $f(x)$ is the density function of X,

- $p(x_i)$ is the probability that the variable X takes on the value x_i.

The first moment is the mean; the second moment is the standard deviation (std); the third moment is the skew; the fourth moment is the kurtosis; see Figure 6.5.

Moments are relatively easy to calculate and can be used for correcting modelling errors. Furthermore, they can be very important during option valuation.

6.3 Some Essential Statistical Tests

We recall only three useful statistical tests.

6.3.1 QQ Test

The Quantile-Quantile test (QQ) is used to check if a random variable has a standard normal distribution. It compares the actual probabilities of the random variable X to the expected probabilities of a normally distributed variable. If the random variable is normally distributed, then all the dot points in Figure 6.6 are on the diagonal line.

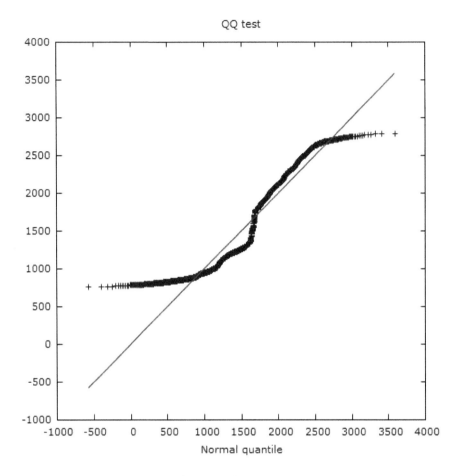

FIGURE 6.6: QQ test

In Figure 6.6 we could distinguish between two behaviours of the random variable. The central values are normal distributed and the behaviour of the tails has to be analysed.

6.3.2   The Autocorrelation Test

The autocorrelation function is used to test normality of a process; actually it is a sequence of autocorrelations. It calculates the various correlation between the steps taken: for total steps, for one-removed steps, for the steps two steps removed, and so on. If indeed the steps are uncorrelated-normally distributed random variables, then all the correlations between steps will be zero. In Figure 6.7 we show the autocorrelation and partial autocorrelation functions of a random walk. We can say that the autocorrelation is the linear dependence of a variable with itself at two points in

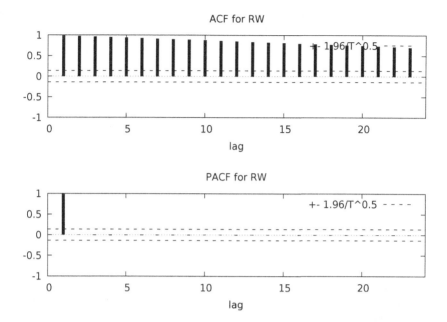

FIGURE 6.7: Autocorrelation and partial autocorrelation functions of a random walk

time. On the contrary, the partial autocorrelation is the autocorrelation between the variable values at two points in time after removing any linear dependence.

6.3.3 R^2

The mean-squared error is the standard deviation of model residuals. The smaller it is, the better our model predicts the actual market data. The R^2 statistic measures how much of the actual uncertainty in the actual data is captured (or explained) by the model being tested. It is measured in percentage terms. If R^2 is 1.0 the model has 100% predictive power.

$$R^2 = 1 - \left(\frac{\text{Mean-squared Error}}{Var(actual\,data)} \right).$$

Bibliography

[1] Juan C. Arismendi, Janis Back, Marcel Prokopczuk, Raphael Paschke, and Markus Rudolf. Seasonal stochastic volatility: Implications for the pricing of commodity options. *Journal of Banking & Finance*, 66:53–65, 2016.

[2] Fred Espen Benth. The stochastic volatility model of Barndorff-Nielsen and Shephard in commodity markets. *Mathematical Finance: An International Journal of Mathematics, Statistics and Financial Economics*, 21(4):595–625, 2011.

[3] Fred Espen Benth and Steen Koekebakker. Stochastic modeling of financial electricity contracts. *Energy Economics*, 30(3):1116–1157, 2008.

[4] Fred Espen Benth, Steen Koekkebakker, and Fridthjof Ollmar. Extracting and applying smooth forward curves from average-based commodity contracts with seasonal variation. *The Journal of Derivatives*, 15(1):52–66, 2007.

[5] Fred Espen Benth, Jurate Saltyte Benth, and Steen Koekebakker. *Stochastic modelling of electricity and related markets*, volume 11. World Scientific, 2008.

[6] William K. Bertram. Analytic solutions for optimal statistical arbitrage trading. *Physica A: Statistical Mechanics and Its Applications*, 389(11):2234–2243, 2010.

[7] Petter Bjerksund, Heine Rasmussen, and Gunnar Stensland. *Valuation and Risk Management in the Norwegian Electricity Market*, pages 167–185. Springer Berlin Heidelberg, 2010.

[8] Fischer Black and Myron Scholes. The pricing of options and corporate liabilities. *Journal of Political Economy*, 81(3):637–654, 1973.

[9] Stanley W. Black. Rational response to shocks in a dynamic model of capital asset pricing. *The American Economic Review*, pages 767–779, 1976.

[10] Oleg Bondarenko. Statistical arbitrage and securities prices. *Review of Financial Studies*, 2003.

[11] Svetlana Borovkova and Helyette Geman. Seasonal and stochastic effects in commodity forward curves. *Review of Derivatives Research*, 9(2):167–186, 2006.

[12] Andrew N. Burgess. A computational methodology for modelling the dynamics of statistical arbitrage models of the FTSE 100. *Ph.D. Thesis, London Business School*, 1999.

[13] Andrew N. Burgess. Statistical arbitrage models of the FTSE 100. *Computational finance, edited by Abu-Mostafa*, 1999.

[14] Roy Cerqueti and Viviana Fanelli. Long memory and crude oil's price predictability. *Annals of Operations Research*, pages 1–12, 2019.

[15] Roy Cerqueti, Viviana Fanelli, and Giulia Rotundo. Long run analysis of crude oil portfolios. *Energy Economics*, 79:183–205, 2019.

[16] Jilong Chen, Christian Ewald, and Ali M. Kutan. Time-dependent volatility in futures contract options. *Investment Analysts Journal*, pages 1–12, 2019.

[17] Rongda Chen and Jianjun Xu. Forecasting volatility and correlation between oil and gold prices using a novel multivariate gas model. *Energy Economics*, 78:379–391, 2019.

[18] Shan Chen and Margaret Insley. Regime switching in stochastic models of commodity prices: An application to an optimal tree harvesting problem. *Journal of Economic Dynamics and Control*, 36(2):201–219, 2012.

[19] Julien Chevallier, Stéphane Goutte, Khaled Guesmi, and Samir Saadi. On the Bitcoin price dynamics: An augmented Markov-Switching model with Lévy jumps. *Working Paper*, 2019.

[20] Les Clewlow and Chris Strickland. Valuing energy options in a one factor model fitted to forward prices. *Available at SSRN 160608*, 1999.

[21] Les Clewlow and Chris Strickland. *Energy Derivatives: Pricing and Risk Management*. Lacima, 2000.

[22] John H. Cochrane. How big is the random walk in gnp? *The Journal of Political Economy*, pages 893–920, 1988.

[23] John C. Cox, Stephen A. Ross, and Mark Rubinstein. Option pricing: A simplified approach. *Journal of financial Economics*, 7(3):229–263, 1979.

[24] Emanuel Derman and Iraj Kani. Riding on a smile. *Risk*, 7(2):32–39, 1994.

[25] Binh Do, Robert Faff, and Kais Hamza. A new approach to modeling and estimation for pairs trading. Monash University, Working Paper, 2006.

[26] Robert J. Elliott, John Van Der Hoek, and William P. Malcolm. Pairs trading. *Quantitative Finance*, 5(3):271–276, 2005.

[27] Alexander Eydeland and Hélyette Geman. Pricing power derivatives. *Risk*, 1998.

[28] Alexander Eydeland and Hélyette Geman. Fundamentals of electricity derivatives. *Energy Modelling and the Management of Uncertainty*, pages 35–43, 1999.

[29] Viviana Fanelli. Commodity-linked arbitrage strategies and portfolio management. *Handbook of Multi-Commodity Markets and Products: Structuring, Trading and Risk Management*, pages 901–938, 2014.

[30] Viviana Fanelli and Anna Katarina Ryden. Pricing a swing contract in a gas sale company. *Economics, Management and Financial Markets*, 13(2):40–55, 2018.

[31] Viviana Fanelli and Maren Diane Schmeck. On the seasonality in the implied volatility of electricity options. *Quantitative Finance*, 19(8):1321–1337, 2019.

[32] Viviana Fanelli, Lucia Maddalena, and Silvana Musti. Modelling electricity futures prices using seasonal path-dependent volatility. *Applied Energy*, 173: 92–102, 2016.

[33] Viviana Fanelli, Lucia Maddalena, and Silvana Musti. Asian options pricing in the day-ahead electricity market. *Sustainable Cities and Society*, 27:196–202, 2016.

[34] Dennis Frestad. Electricity forward return correlations. *Working Paper*, 2007.

[35] Dennis Frestad. Common and unique factors influencing daily swap returns in the Nordic electricity market, 1997–2005. *Energy Economics*, 30(3):1081–1097, 2008.

[36] Gianluca Fusai, Marina Marena, and Andrea Roncoroni. Analytical pricing of discretely monitored asian-style options: Theory and application to commodity markets. *Journal of Banking & Finance*, 32(10):2033–2045, 2008.

[37] Hélyette Geman. *Risk management in commodity markets: From shipping to agriculturals and energy*, volume 445. John Wiley & Sons, 2009.

[38] Hélyette Geman and Andrea Roncoroni. Understanding the fine structure of electricity prices. *The Journal of Business*, 79(3):1225–1261, 2006.

[39] Rajna Gibson and Eduardo S. Schwartz. Stochastic convenience yield and the pricing of oil contingent claims. *The Journal of Finance*, 45(3):959–976, 1990.

[40] Abebe Hailemariam and Russell Smyth. What drives volatility in natural gas prices? *Energy Economics*, 2019.

[41] David Heath, Robert Jarrow, and Andrew Morton. Bond pricing and the term structure of interest rates: A new methodology for contingent claims valuation. *Econometrica: Journal of the Econometric Society*, pages 77–105, 1992.

[42] Harold Hotelling. Analysis of a complex of statistical variables into principal components. *Journal of Educational Psychology*, 24(6):417, 1933.

[43] Harold Hotelling. Relations between two sets of variates. In *Breakthroughs in statistics*, pages 162–190. Springer, 1992.

[44] Ronald Huisman and Ronald Mahieu. Regime jumps in electricity prices. *Energy Economics*, (25):425–434, 2003.

[45] Kiyosi Itô. *On stochastic differential equations*, volume 4. American Mathematical Soc., 1951.

[46] Joanna Janczura and Rafal Weron. An empirical comparison of alternate regime-switching models for electricity spot prices. *Energy Economics*, 32(5): 1059–1073, 2010.

[47] Vincent Kaminski. *Managing energy price risk: The new challenges and solutions*. Risk Books, 2004.

[48] Rüdiger Kiesel, Gero Schindlmayr, and Reik H Börger. A two-factor model for the electricity forward market. *Quantitative Finance*, 9(3):279–287, 2009.

[49] Peter E. Kloeden and Eckhard Platen. *Numerical solution of stochastic differential equations*. Springer-Verlag, 1999.

[50] Bingxin Li. Pricing dynamics of natural gas futures. *Energy Economics*, 78: 91–108, 2019.

[51] Andrew W. Lo and A. Craig MacKinlay. Stock market prices do not follow random walks: Evidence from a simple specification test. *Review of Financial Studies*, 1(1):41–66, 1988.

[52] Julio J. Lucia and Eduardo S. Schwartz. Electricity prices and power derivatives: Evidence from the nordic power exchange. *Review of Derivarives Research*, 5(1):5–50, 2002.

[53] Carlo Mari. Regime-switching characterization of electricity price dynamics. *Physica A: Statistical Mechanics and Its Applications*, (371):552–564, 2006.

[54] Robert L. McDonald. *Derivatives markets: Pearson New International Edition*. Pearson Higher Ed, 2013.

[55] Robert C. Merton. Option pricing when underlying stock returns are continuous. *Journal of Financial Economics*, 3:125–144, 1976.

[56] Dragana Pilipovic. *Energy risk: Valuing and managing energy derivatives*. McGraw Hill Professional, 2007.

[57] Andrew Pole. *Statistical arbitrage*. Wiley Finance, 2007.

[58] Maren Diane Schmeck. Pricing options on forwards in energy markets: the role of mean reversion's speed. *International Journal of Theoretical and Applied Finance*, 19(08):1650053, 2016.

[59] Eduardo Schwartz and James E. Smith. Short-term variations and long-term dynamics in commodity prices. *Management Science*, 46(7):893–911, 2000.

[60] Eduardo S. Schwartz. The stochastic behavior of commodity prices: Implications for valuation and hedging. *The Journal of Finance*, 52(3):923–973, 1997.

[61] Carsten Sørensen. Modeling seasonality in agricultural commodity futures. *Journal of Futures Markets: Futures, Options, and Other Derivative Products*, 22(5):393–426, 2002.

[62] Anders B. Trolle and Eduardo S. Schwartz. Unspanned stochastic volatility and the pricing of commodity derivatives. *The Review of Financial Studies*, 22(11): 4423–4461, 2009.

[63] Rafał Weron. Market price of risk implied by Asian-style electricity options and futures. *Energy Economics*, 30(3):1098–1115, 2008.

[64] Rafał Weron, Michael Bierbrauer, and Stefan Trück. Modeling electricity prices: Jump diffusion and regime switching. *Physica A: Statistical Mechanics and Its Applications*, 336(1-2):39–48, 2004.

Index

Printed in the United States
by Baker & Taylor Publisher Services